Johannes Fahrentrapp

Fire Blight Resistance of Malus x robusta 5

Johannes Fahrentrapp

Fire Blight Resistance of Malus x robusta 5

Südwestdeutscher Verlag für Hochschulschriften

Impressum / Imprint
Bibliografische Information der Deutschen Nationalbibliothek: Die Deutsche Nationalbibliothek verzeichnet diese Publikation in der Deutschen Nationalbibliografie; detaillierte bibliografische Daten sind im Internet über http://dnb.d-nb.de abrufbar.
Alle in diesem Buch genannten Marken und Produktnamen unterliegen warenzeichen-, marken- oder patentrechtlichem Schutz bzw. sind Warenzeichen oder eingetragene Warenzeichen der jeweiligen Inhaber. Die Wiedergabe von Marken, Produktnamen, Gebrauchsnamen, Handelsnamen, Warenbezeichnungen u.s.w. in diesem Werk berechtigt auch ohne besondere Kennzeichnung nicht zu der Annahme, dass solche Namen im Sinne der Warenzeichen- und Markenschutzgesetzgebung als frei zu betrachten wären und daher von jedermann benutzt werden dürften.

Bibliographic information published by the Deutsche Nationalbibliothek: The Deutsche Nationalbibliothek lists this publication in the Deutsche Nationalbibliografie; detailed bibliographic data are available in the Internet at http://dnb.d-nb.de.
Any brand names and product names mentioned in this book are subject to trademark, brand or patent protection and are trademarks or registered trademarks of their respective holders. The use of brand names, product names, common names, trade names, product descriptions etc. even without a particular marking in this works is in no way to be construed to mean that such names may be regarded as unrestricted in respect of trademark and brand protection legislation and could thus be used by anyone.

Coverbild / Cover image: www.ingimage.com

Verlag / Publisher:
Südwestdeutscher Verlag für Hochschulschriften
ist ein Imprint der / is a trademark of
AV Akademikerverlag GmbH & Co. KG
Heinrich-Böcking-Str. 6-8, 66121 Saarbrücken, Deutschland / Germany
Email: info@svh-verlag.de

Herstellung: siehe letzte Seite /
Printed at: see last page
ISBN: 978-3-8381-3091-0

Zugl. / Approved by: Zürich, ETH, Dissertation, 2012

Nothing is stronger than an idea whose time has come.

Victor Hugo

Abstract

Fire Blight (FB) is the most important bacterial disease in apple (*Malus × domestica*) and pear (*Pyrus communis*) production. The application of resistant cultivars would be an additional measure against FB limiting the dependence on antibiotic treatments and the dormant risk for producers to lose their only effective control measure. However, today there is no popular FB resistant apple variety available. Strong quantitative trait loci (QTLs) for resistance were identified e.g. in the wild apple *'Malus × robusta* 5' (MR5). The present work aimed firstly to provide usable molecular markers tightly linked to the resistance locus of MR5 and secondly to isolate and characterize the source of MR5 resistance against *Erwinia amylovora.* 14 new molecular markers were developed and mapped to the top of linkage group (LG) 3 of MR5 containing the resistance locus. Together, they defined the distal 1.5 cM of LG 3 of MR5. After phenotyping of recombinant individuals of an expanded population of more than 2000 individuals segregating for MR5 resistance, the locus of the resistance trait was flanked by two markers defining an interval of 0.23 cM. By means of *chromosome landing* two bacterial artificial chromosomes (BACs) harboring the genomic DNA of MR5 were identified, one of which was in coupling and one in repulsion to the alleles of the flanking markers. Using differently trained protein prediction software (FGENESH) 25-47 open reading frames (ORFs) were predicted on the assembled sequences derived from the resistance carrying BAC. The predicted mRNAs were annotated and compared to the constitutional transcriptome of MR5 to evaluate the prediction algorithms. The annotation and subsequent motif analysis led to the identification of one candidate resistance (*R*) gene named *FB_MR5.* We investigated the candidate resistance gene, which was demonstrated to be constitutionally expressed in MR5 as well as in resistant progenies of MR5, but not in susceptible cultivars ('Idared', 'La Flamboyante', 'Gala') and susceptible progenies of MR5. *FB_MR5* was cloned and resequenced. FB_MR5 was classified *in silico* as CC-NBS-LRR protein (CNL, coiled coil nucleotide binding site leucine rich repeat). All investigated CNLs giving resistance towards bacterial diseases function in accordance to the 'decoy model' or 'guard model' which postulate the interaction of (1) the R protein of the host, which recognizes conformational changes or degradation of (2) the 'decoy'/'guardee' protein interacting with (3) the bacterial effector (Avr protein). Taking into account the recently published findings on host-microbe interactions in other plants and the presence and absence, respectively, of transcribed putative decoys/guardees in MR5 transcriptome as well as the presence and absence, respectively, of Avr genes in *E. amylovora*, we hypothesized a homologous mode of function of FB_MR5 to RPS2 of *Arabidopsis thaliana*. Finally, to give evidence to the functionality of FB_MR5 acting as *R* gene against FB, a complementation assay into susceptible cultivar 'Gala' was appended as preliminary results in this thesis. Two constructs comprising *FB_MR5* under the control of CaMV 35s promoter and its own promoter, respectively, were cloned into 'Gala' via *Agrobacterium tumefaciens*. 24 newly developed transgenic shoots are growing to a usable size for micro grafting on M9 rootstocks and will later be inoculated with *E. amylovora*.

Zusammenfassung

Feuerbrand (FB) ist die bedeutendste bakterielle Krankheit bei der Apfel- (*Malus* × *domestica*) und Birnenproduktion (*Pyrus communis*). Mit dem Anbau resistenter Apfelsorten wäre eine zusätzliche Maßnahme gegen FB verfügbar. Dadurch würden die Produzenten unabhängiger vom Antibiotikagebrauch und das potenzielle Risiko, diese einzige effektive Kontrollmaßnahme zu verlieren, wäre geringer. Gegenwärtig ist keine beliebte FB-resistente Apfelsorte erhältlich. Starke *Quantitative Trait Loci* (QTL; genomische Loki) für Resistenz wurden zum Beispiel im Wildapfel ‚*Malus* × *robusta* 5' (MR5) identifiziert. Die vorliegende Arbeit hatte zum Ziel erstens molekulare Marker, die eng gekoppelt sind an den Resistenzlokus von MR5, zu entwickeln, und zweitens die Resistenzquelle von MR5 gegen *Erwinia amylovora* zu isolieren und zu charakterisieren. 14 neue molekulare Marker wurden entwickelt und an die Spitze von Kopplungsgruppe (*linkage group*, LG) 3 von MR5 kartiert, die den Resistenzlokus beinhaltet. Zusammen definierten sie die distalen 1,5 cM der Spitze von LG 3 von MR5. Nach der Phänotypisierung rekombinanter Individuen einer vergrößerten Population von mehr als 2000 Individuen, die bezüglich der MR5 Resistenz segregieren, konnte der Resistenzlokus mit zwei Markern flankiert werden, die ein Intervall von 0,23 cM definieren. An Hand von *chromosome landing* konnten zwei BACs (*bacterial artificial chromosomes*) gefunden werden, die DNA Fragmente von MR5 in Kopplung bzw. in Repulsion bezüglich der Allele der flankierenden Marker tragen. Mit unterschiedlich geschulter Software (FGENESH) wurden auf der Sequenz des BAC, das die Resistenz trägt, 25-47 *open reading frames* (ORFs) vorhergesagt. Die vorhergesagten mRNAs wurden annotiert und mit dem konstitutionellen Transkriptom von MR5 verglichen um die Algorithmen der Prädiktionssoftware zu beurteilen. Die Annotation und anschließende Motivanalysen führten zur Identifikation eines Kandidatenresistenzgens, das *FB_MR5* genannt wurde. Wir untersuchten das Kandidatenresistenzgen, das konstitutionell in MR5 sowie in resistenten Nachkommen von MR5 nicht aber in anfälligen Sorten (‚Idared', ‚La Flamboyante', ‚Gala') sowie in anfälligen Nachkommen von MR5, exprimiert wurde. *FB_MR5* wurde kloniert und sequenziert. FB_MR5 wurde *in silico* als CC-NBS-LRR Protein (CNL, *coiled coil nucleotide binding site leucine rich repeat*) klassifiziert. Alle erforschten CNLs, die Resistenz gegen bakterielle Krankheiten vermitteln, funktionieren nach dem ‚Lockvogel-' oder ‚Wächter-Modell' (‚decoy-'/‚guard-model'), das mindestens drei zusammenarbeitende Genprodukte postuliert: (1) das Resistenzprotein des Wirts erkennt die Konformationsänderungen oder den Abbau des (2) Lockvogel/*Guardee*-Proteins während der Interaktion mit dem bakteriellen (3) Effektorprotein (Avirulenz (Avr) Protein). Unter Berücksichtigung der aktuellen Forschungsergebnisse bezüglich der Wirt-Schadorganismus-Interaktion bei anderen Pflanzen und des Vorhanden- oder Nichtvorhandenseins des Lockvogel/*Guardee*-Proteins im Transkriptom von MR5 sowie des Vorhanden- oder Nichtvorhandenseins des Avr-Proteins in *E. amylovora*, formulierten wir die Hypothese, dass FB_MR5 homolog zu RPS5 von *Arabidopsis thaliana* funktioniert. Schlussendlich, um die Funktionalität von *FB_MR5* als Resistenzgen gegen FB zu beweisen, wurde ein Komplemetationsversuch bei der anfälligen Sorte ‚Gala' durchgeführt, dessen vorläufige Ergebnisse dieser Doktorarbeit hinzugefügt wurden. Zwei Konstrukte mit *FB_MR5* unter der Kontrolle seines eigenen Promotors bzw. des CaMV 35S-Promotors wurden mittels *Agrobacterium tumefaciens* in ‚Gala' kloniert. 24 neu entwickelte transgene Triebe sind gewachsen und werden nach Erreichen der notwendigen Größe auf M9-Unterlagen mikro-veredelt und später mit *E. amylovora* inokuliert.

Table of Contents

Chapter 1

General Introduction

Preface

Since the bacterium *Erwinia amylovora* causing the disease Fire Blight spread from North America to almost all over the world and reaching Europe in 1958, the control of this disease has become an increasing task for apple and pear producers. In 1989 the causing bacteria was observed for the first time in Switzerland at Stein am Rhein and colonized most of the country in the following years leading to decreasing yields of the affected crops. The damage of apple and pear trees in Switzerland caused by *E. amylovora* infections reached its maximum in 2007. Consequently new strategies to control the disease should be developed and established. Therefore scientific efforts in basic and applied research on fire blight, its causing bacterium and resistance of apple and pear trees against the disease was supported by several governmental research programs such as ZUEFOS (Züchtung feuerbrandtoleranter Obstsorten, breeding of fire blight tolerant fruit varieties) from Federal Office for Agriculture (FOAG, Switzerland) and the D-A-CH project (German-Austrian-Swiss project) with the title 'Identification, cloning and functional characterization of genes related to fire blight'. ZUEFOS was setup in four modules one of which aimed at the identification of molecular markers tightly linked to fire blight resistance of *Malus* × *robusta* 5 and to investigate the resistance locus. The D-A-CH project aimed at the in depth analysis and characterization of the same resistance locus on a molecular level. In the presented doctoral thesis the execution and results of parts of these projects are described which comprise the identification of a candidate fire blight resistance gene, referred to as *FB_MR5*, its expression as well as the analysis of the surrounding genetic region.

Plant Resistance – Molecular Plant-Microbe Interaction – Resistance Models

In general, plants are resistant against most insects and pathogenic organisms. This is referred to as 'nonhost resistance' and the plant as well as the pathogen is termed as nonhost plant and pest/pathogen, respectively (Hans 2003). For example, plants evolved preformed defense mechanisms against insects and pathogens such as the cuticular waxes, the cell wall and the lignification of the latter. If pathogenic organisms are able to colonize or even enter their hosts, plants react in specific manner depending to the invader. For example, chewing insects induce in plants the production of protease inhibitors, alkaloids and the release of volatiles that attract other insects feeding on the pathogenic insect (Kessler and Baldwin 2002). Furthermore, peptides like systemin are produced by plants after wounding which elicit plant defense (Pearce et al. 1991; Scheer and Ryan 2002). Nematodes, fungi, oomycetes, viruses and bacteria attack their host plants mostly through natural openings, wounds or with the use of cell wall degrading enzymes, which can be recognized by plants stimulating defense reactions. The so-called 'basal immune system' of plants reacts after recognition of pathogenic molecules, referred to as elicitors, in the so-called 'basal defense' or as pathogen triggered immunity (PTI) (Jones and Dangl 2006). The elicitors also called pathogen-associated molecular patterns (PAMPs) or microbe-associated molecular patterns (MAMPs) are molecules of pathogens which evolve slowly, which are highly conserved and which are mostly essentially for pathogen viability. MAMPs are molecules such as bacterial flagellins, lipopolysaccharides or elongation factor-Tu, fungal chitin, or oomycete Pep-13 (summarized by Bent and Mackey 2007). MAMPs like the bacterial flaggelin flg22 of *Pseudomonas syringae* are recognized by a plant pattern recognition receptors (PRRs) like FLS2 of *Arabidopsis thaliana* (Gómez-Gómez and Boller 2000; Zipfel 2008). PRRs, preformed phytoalexins and physical barriers lead to additive resistance, a so-called quantitative or horizontal resistance described by

Van der Plank (1963). PTI is modulated or repressed from pathogen side by the injection of proteins referred to as effectors (Nomura et al. 2005). Some of those effectors are also called avirulence (Avr) proteins, if they are recognized directly or indirectly by the plant through a specific protein encoded in a so called 'resistance' (*R*) gene. Upon recognition the plant activates a defense cascade which leads to incompatibility (avirulence of the pathogen). Bacteria inject the effectors typically through a type III secretion system (T3SS) directly into the cytoplasma of plants cells (Martin et al. 2003) but also the secretion systems type IV and VI of bacteria are able to transfer effectors to eukaryotic cells (Hayes et al. 2010). Recognition of effectors by R gene products lead to an effector triggered immunity (ETI) often in form of a hypersensitive response (HR) in plant tissue. HR is the *R* gene induced programmed cell death around pathogen infection side (Heath 2000). The resistance given by single *R* genes is usually pathogen race specific and referred to as major or vertical resistance (Van der Plank 1963). The mode of the recognition of effectors by R gene products was simply explained by the direct gene-for-gene interaction (Flor 1971). This interaction concept was evidenced in yeast-2-hybrid experiments showing the interaction of the rice Pi-ta resistance protein and AVR-Pita of *Magnaporthe grisea* causing rice blast disease (Jia et al. 2000); as well as the direct interaction of the flax (*Linum usitatissimum*) L5, L6 and L7 resistance gene products with AvrL567 of the flax rust fungus *Melampsora lini* (Dodds et al. 2006). The gene-for-gene concept was supplemented by the guard-model (Van der Biezen and Jones 1998; Dangl and Jones 2001) to explain interactions which were found to be indirect. The guard-model expands the gene-for-gene concept describing the recognition of Avr by a 'guardee' being guarded by an *R* gene (the 'guard'). The guardee was defined as target for the corresponding virulence factor. This was demonstrated for instance for *P. syringae*'s effector AvrPphB, which cleaves the protein kinase PBS1 of *A. thaliana*, whereas the cleavage is detected by the guarding R protein RPS5 (Caplan et al. 2008). Another example formerly described by the guard model is the cleavage of Pto (*Solanum lycopersicum*) guarded by Prf whereas Pto is cleaved by pathogen injected AvrPto/AvrPtoB (*P. syringae* pv. *tomato*) (Van der Biezen and Jones 1998). However, it was demonstrated that the main targets of the effector AvrPto were the kinase domains of the receptor-like kinases CERK1, BAK1, EFR1 and FLS2 instead of Pto (summarized in Moffett 2009). To explain such cases the decoy model was introduced (Van der Hoorn and Kamoun 2008). The decoy takes over the position of the guardee and instead of the guardee the decoy does not enhance fitness of the pathogen in absence of the *R* gene. The historic definition of an *R* gene was summarized by Bent (2007) to "polymorphic plant genes that control gene-for-gene disease resistance" which doesn't match all the different known *R* genes such as *Prf*, which is not polymorphic but its decoy *Pto*. The difference of the defining characteristics of decoy or guardee towards *R* gene is also not clearly determined. For example the kinase Pto was formerly described as R protein, but since it was found to be monitored by Prf it has the position of the decoy and Prf is the R protein but Pto remains an R protein for many researchers (Bent and Mackey 2007).

Plant R Proteins – Their Structure and Function

Most R proteins comprise a nucleotide binding site (NBS) and a leucine rich repeat domain (LRR) (Ellis et al. 2000) but also other proteins were assigned as R proteins such as kinases (Pto) or the receptor like kinases Xa21.

The NBS domains belong to the STAND superfamily of ATPases (Signal Transduction ATPases with Numerous Domains) and are capable to bind ADP and hydrolyze ATP (DeYoung and Innes 2006). The NBS can be subdivided into the NB domain followed by the so-called ARC 1 and ARC 2 domains which comprise the conserved motifs hhGRExE (where h represents a hydrophobic and x any residue), Walker A (synonym: P-loop as well as kinase 1a), Walker B (synonym: kinase 2), RNBS-A, -B, -C and -D, GLPL (synonym: GxP), Motif VII, VIII and X as well as MHD (Figure 1.1) (van Ooijen et al. 2007). Most of these motifs are indispensable for R protein function (Takken et al. 2006; Van Ooijen et al. 2008).

The LRR proteins are built of repeated amino acid (aa) elements one of which has the minimal general consensus LxxLxL (Kajava and Kobe 2002) with typically 21-25 aa residues. LRR elements forming the LRR protein are several times repeated for example 11 times in an apple NBS-LRR coding for a putative resistance gene against *E. amylovora* (Parravicini et al. 2011) and 27 times in L6, an R protein of *L. usitatissimum* detecting *M. lini* (Jones and Jones 1997). LRRs form a concave structure as shown in the first plant LRR of polygalacturonase-inhibiting protein of *Phaseolus vulgaris* (Di Matteo et al. 2003) and depending on the number of repeats this structure can be enlarged to a horse shoe shaped protein (Kobe and Deisenhofer 1994).

The NBS-LRR proteins can be subdivided into two main classes defined by their N-terminus which is either formed by a TIR (Toll/interleukin 1 receptor) domain or a coiled-coil (CC) region which are abbreviated as TNL and CNL, respectively. The Toll/interleukin-1 receptor (TIR) domain is characterized by three conserved motifs (Slack et al 2000) and is mostly present in dicots with only one example in monocots (Meyers et al. 2002). Homologue TIR domains are also found in toll-like receptors (TLR) of *Drosophila*, activating antifungal defense response (Kopp and Medzhitov 1999), and in interleukin 1 (IL-1) receptors of humans (Ausubel 2005) also active in innate immune response. The role of TIR in the TNL complex in plants was thought to signal in pathways towards apoptosis but it was also shown, that it directly causes cell death (Swiderski et al. 2009). CC domains are formed by a minimum of 2 alpha-helices that wind around each other in a supercoiled structure (Gruber et al. 2006). CC domains of CNLs mostly possess two coiled coil structures, that are separated by the conserved aa motif 'EDVID' (Collier and Moffett 2009). The CC domain was shown to interact with proteins. For example the CC domain of Rx, a resistance protein of potato against potato virus X, interacts with RanGAP2 a protein of potato (Sacco et al. 2007; Rairdan et al. 2008), and the CC region of RPM1 of *A. thaliana*, conferring resistance to *P. syringae*, interacts with RIN4 of *A. thaliana* (Mackey et al. 2002).

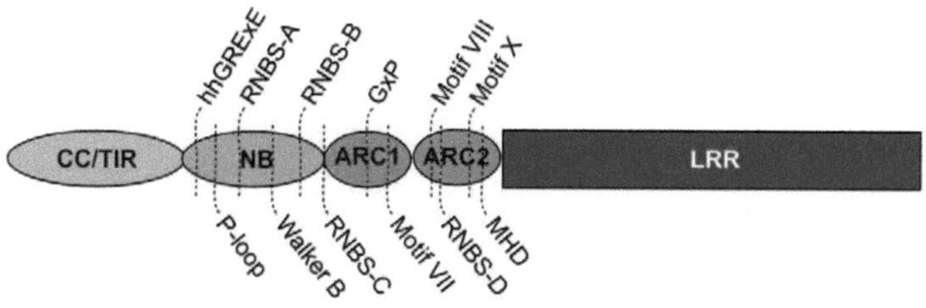

Figure 1.1 Conserved motifs of the NBS (NB-ARC) domains. The NB-ARC domain belongs to the STAND family of ATPases and can be subdivided into the NB, ARC1 and ARC2 domain comprising conserved motifs as stated in the figure (from van Ooijen et al. 2007).

The mode of function of the whole protein complex CC/TIR-NBS-LRR depends on the interaction of the involved domains. The N-terminal domain (CC or TIR) seems to be responsible for downstream signaling whereas the LRR is most important for effector recognition (Martin et al. 2003). As shown in Figure 1.2 (Takken and Tameling 2009) ADP is bound between the NBS, ARC1 and ARC2 domains in a highly stable and autoinhibited resting state ('standby') of the protein (Tameling et al. 2006). During the interaction of an effector with the LRR, ADP is released, and a conformational change to a more open conformation ('intermediate' state) is introduced which allows binding of ATP (Tameling et al. 2006). This leads to another conformational change altering the interaction between NBS and the N-terminal TIR/CC domains allowing the NBS to initiate plant defense cascade ('on' state) (Takken and Tameling 2009). After ATP hydrolysis the complex rearranges to his resting state.

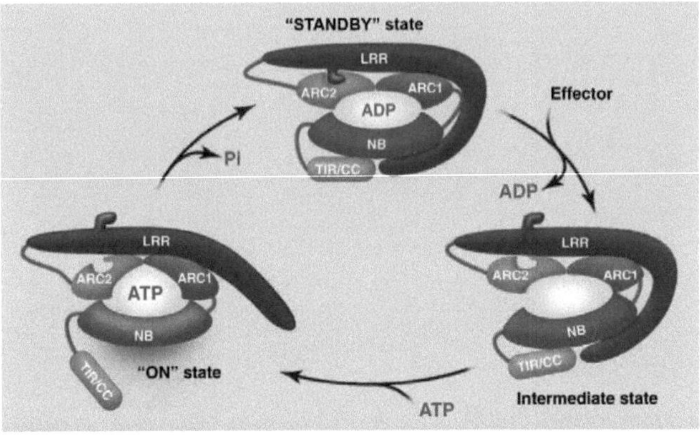

Figure 1.2 Model for the mode of function of CC/TIR-NBS-LRRs. The autoinhibited resting state ('standby') of the protein is present when ADP is bound to the central NBS (NB) domain. If a pathogen derived effector interacts with the LRR, ADP is released changing the conformation to an 'intermediate' state which then allows binding of ATP. The following 'on' state of the protein alters the interaction between NBS and the N-terminal TIR/CC domains allowing the NBS to initiate the plant defense cascade. If ATP is hydrolyzed to ADP the protein rearranges to his resting state. Figure modified from Takken and Tameling (2009).

Durability of *R* Genes

R genes are effective as long as the pathogen cannot establish a mechanism to escape the recognition of R proteins. Hence, the durability of *R* genes depends mainly on the evolutionary potential of the pathogen and the mode of deployment of the resistant cultivar. A prediction concept was introduced classifying pathogens into risk classes depending on their mode of reproduction, mutation rate, populations size, diversity etc. which classified *E. amylovora* in a lower risk class than other bacterial plant pathogens such as *Xanthomonas compestris* pv. *vesicatoria*, *X. malvacearum* and *X. oryzae* pv. *oryzae* (McDonald and Linde 2002). But these predictions did not always correlate to what happens in nature as shown e.g. for Xa-4 against race 2 of *X. oryzae* pv. *oryza*e which was broken within only half of the predicted years (summarized in McDonald and Linde 2002). Additional other *R* genes remain effective despite they are widely used such as the resistance of the rice cultivar Nonken 58 against *X. oryzae* pv. *oryzae* which lasts more than 25 years (Bonman et al. 1992). The *R* gene *Vf* against apple scab caused by *Venturia ineaqualis* was broken after intensive use in apple breeding and may serve as an example for *R* gene breakdown in perennial crops (Gessler and Pertot 2011). In general the time frame until breakdown of deployed *R* genes can be extended by pyramiding different *R* genes in one cultivar as well as crop rotation and mixed cultures limiting the selection pressure and therefore obviating the increase to high frequency of the selected resistant pathogen strains.

Bacterial Effectors

Effectors found in bacteria are mostly injected into host cells via the T3SS which are therefore referred to as type 3 (T3) effectors. Some of these effectors were coded by so-called *hrp* (HR and pathogenicity) genes. The T3SS was formerly known as *hrp*-pilus because it is formed by a set of proteins encoded by some of these *hrp* genes (Bent and Mackey 2007). In a recent review on T3 effectors 88 effectors of plant diseasing bacteria are listed as proven effectors (Dean 2011). The same review accentuates the high diversity of T3 effectors which are usually built up in a segmental chain of functional domains. The N-terminal domain always comprises a 15-25 aa signal sequence obligatory for T3SS secretion; the subsequent series of domains are not homolog between the different effectors (Dean 2011).

In *E. amylovora* 17 proteins are described to be delivered via T3SS: HrpJ, HrpQ, HrpA, HrpN, HrpW, HopPtoC, AvrRpt2$_{EA}$*, DspA*, DspE*, Eop1*, Eop2 (HopAK1$_{Ea}$), Eop3* (HopX1$_{Ea}$), EopB* and OrfB* whereas only the ones marked with a * may be recognized as effectors (Kim and Beer 2000; Dean 2011; Khan et al. 2011; Bocsanczy et al. 2012). The homolog of

AvrRpt2 of *P. syringae* in *E. amylovora*, $AvrRpt2_{EA}$, was demonstrated to elicit HR in *A. thaliana* when inoculated with $avrRpt2_{EA}$ expressing transgene *P. syringae* (Zhao et al. 2006). The corresponding R protein RPS2, a CNL, and its decoy/guardee RIN4 of *A. thaliana* are already investigated in depth (Axtell and Staskawicz 2003).

The Apple

Apple plants are very well adapted species of the temperate fruit crops growing from 35-50 ° latitude and in heights with temperatures down to -40°C as well as tropic regions where two harvests a year are produced (Forsline et al. 2010; Ignatov and Bodishevskaya 2011). The cultivated apple trees were known to belong to the *Rosaceae* subfamily *Maloideae* (synonym: *Pomoideae*) named scientifically *Malus* × *domestica* Borkh. and due to its high level self-incompatibility the bred cultivars are interspecific hybridizations (Korban 1986). Recently a new nomenclature based on molecular markers classified the genus *Malus* as member of the family *Rosaceae*, subfamily *Spiroideae* and subtribe *Pyrinae* (former *Maloideae*), which includes also *Pyrus*, *Sorbus*, *Cotoneaster* and others (Campbell et al. 2007; Potter et al. 2007). Apples are growing on 3-12m tall shrub or trees and have a generation time between 3-8 years. The flowers carry both sexes and apple cultivars are usually self-incompatible what is determined by the S-alleles of the self-incompatibility locus (Juniper and Mabberley 2006). Ripening of apples ranges from 70 days to 180 days after bloom depending on the cultivar (Ignatov and Bodishevskaya 2011). The apple's origin is believed to be localized in Central Asia and the most important progenitor is described with *M. sieversii* Lebed. (Forsline et al. 2010). From the wild apple forests in Central Asia apple trees and seeds were dispersed to China and to Europe via the "Silk Route".

The beginning of apple cultivation dates back at least 3000 years to Greece and Persia and already in the third century BC Theophrastus described the grafting techniques. During the 19th century more than 7,000 cultivars were described (Ragan 1905) increasing to a number of more than 10,000 cultivars in the end of the last century (Janick et al. 1996). Today more than 71 million tons of apples (FAO 2009), which are mostly successions of *M.* × *domestica* Borkh., are produced worldwide making apple one of the most cropped fruits with still increasing potential. Most apples are produced in China (32 Mt (Mega tons)) followed by the US (4.5 Mt), Turkey (2.8 Mt) and Poland (2.6 Mt) (FAO 2009). Switzerland produces 0.25 Mt (FAO 2009). The most sold apple cultivars in Europe are 'Golden Delicious' with about 30% market share in the period from 2007 to 2011 followed by 'Gala' with 9% (www.statista.com); in the US fresh 'Gala' reaches 22% market share followed by 'Red Delicious' (16%) and 'Fuji' (13%) (www.statista.com) and the market in China, the world leading country in terms of apple production, is dominated by 'Fuji' with 60% market share (Heng et al. 2008). 38% of the Swiss apple production area is used for cultivation of 'Golden Delicious' and 'Gala' (Bravin and Kilchenmann 2010). The concentration to a very small number of grown cultivars resulted in new tasks for breeders: The need arises for pest and disease resistant traits as well as a stronger stress tolerance (Ignatov and Bodishevskaya 2011).

The genome of *M.* × *domestica* was estimated on the basis of the partial genome sequence (approximately 67% were sequence and assembled) of the cultivar 'Golden Delicious' to a size of 742.3 Mbp (mega base pairs) (Velasco et al. 2010). With exceptions most *Malus* species have a diploid genome with 17 chromosomes (2n = 34) but also triploid cultivars (e.g. 'Jonagold') and tetraploids (e.g. 'Yellow Transparent') are known (Janick et al. 1996). All chromosomes were mapped in a genetic linkage reference map with 840 molecular markers spanning between 1,140-1,450 cM (Liebhard et al. 2003). The genetic map of apple was enhanced with additional markers (n=1,046) leading to an average marker density of 1.1 cM/marker and a total length of 1,032 cM (N'Diaye et al. 2008). According to these numbers of total cM, in average 0.5-0.7 kbp would correlate to one cM.

Apple Diseases

Apple plants are usually planted in monocultures which increase the risk of serious pests and diseases. The most important diseases in apple production are apple scab caused by the fungus *Venturia inequalis*, powdery mildew caused by *Podosphera leucotricha* and fire blight caused by the bacterium *E. amylovora*, but also other diseases are known like crown rot (*Phytophthora cactorum*), cedar apple rost (*Gymnosporangium juniper-virginianae*), Valsa canker (*Valsa ceratosperma*) and Nectria canker (*Nectria galligena*). The important pests are wooly apple aphids (*Ericoma lanigerum*), rosy apple aphids (*Dysaphis plantaginea*) and rosy curling aphids (*D. devecta*) (summarized in Way 1991).

Fire Blight

Fire blight is besides apple scab the disease with the highest impact on apple production. Fire blight is caused by the Gram-negative bacterium *Erwinia amylovora* (see below) which enters its host through wounds or natural openings like nectarthodes. It is dispersed by rain splash, wind and pollinators. Once entered the host it multiplies and moves intercellularly downwards. Visible symptoms are necrotized tissues, a milky to orange bacterial ooze (Figure 1.3A) and, after the bacteria reached the shoot tip, the characteristic Sheppard's crook (Figure 1.3B). Later cankers on woody plant tissues become visible (whole disease cycle: Figure 1.4). The fastness of colonization causes that the leaves remain on the tree and turn brown or black which gave the disease its name 'fire blight'. The fire blight symptoms were first recognized in 1780 in New York State and *E. amylovora* was identified as causal agent in 1902; it invaded Japan in 1903,

reached New Zealand in the 1920ies and spread to Europe in 1958 (reviewed in Peil et al. 2009). In the year 2006 46 countries reported the presence of *E. amylovora* (Van der Zwet 2006). In 1989 *E. amylovora* was detected for the first time in Switzerland (Agroscope-Changins/Wädenswil 2012) initiating the spread of the disease across the whole country with greatest losses in 2007, when apple and pear trees on more than 100 ha acreage were destroyed and more than 45,000 standard trees were infected (Holliger et al. 2007). The financial impact of measures against fire blight in Switzerland between the year 2000 and 2008 was tabulated to 45 million Swiss Francs (Parravicini 2010). Other sources gave numbers of up to 50 million Swiss Francs only for the major outbreak in 2007 (NZZ 2007).
The disease can be partially controlled by chemicals, copper compounds and bio control products (Ngugi et al. 2011). Best control results were obtained by the application of antibiotics such as streptomycin, tetracyclins (Ngugi et al. 2011) and kasugamycin (McGhee and Sundin 2010). In many countries the application of antibiotics in agriculture is prohibited or strongly restricted. With forecasting models such as MaryBlyt™ (www.agrometeo.ch, Agroscope-Changins/Wädenswil, www.caf.wvu.edu/kearneysville/maryblyt/) days of high infection risk are predicted based mainly on weather conditions. The data are accessible to farmers giving the time slots when antimicrobial compounds should be applied best. An additional possibility to control fire blight would be to crop resistant cultivars. Today there is no high quality resistant apple or pear cultivar on the market but several potential sources of resistances are known. Major quantitative trait loci (QTLs) linked to fire blight resistance in apple were found on *M.* × *domestica* cultivars 'Florina' (linkage group (LG) 10) and 'Fiesta' (LG 7), *M. floribunda* 821 (LG 12), the ornamental cultivar 'Evereste' (LG 12) and the wild apple *M.* × *robusta* 5 (LG 3) (Khan et al. 2011). The strongest QTL against FB is the one of *M.* × *robusta* 5 (MR5) on LG 3 explaining up to 80% of the phenotypic variation between susceptibility (100% PLL) and resistance (0% PLL) towards FB in the cross 'Idared' × MR5 (Peil et al. 2007). The QTL of 'Evereste' which

explains up to 68% of the phenotypic variation (Durel et al. 2009) was explored more in detail and several candidate *R* genes were found and sequenced (Parravicini et al. 2011). Two candidates are short-listed. One shows homology to the kinase coding gene *Pto* of tomato (*Solanum lycopersicum*) the other codes for a putative CNL. Both are processed further in complementation assays with results expected soon (C Gessler, personal communication).

Figure 1.3 Fire blight symptoms. A, bacterial ooze produced on apple plants by *E. amylovora* (Ea222_JKI) 21 dpi; B, artificially *E. amylovora* (Ea222_JKI) infected shoots showing resistant (left) and susceptible (right) phenotype, respectively. The photograph was taken at 21 dpi. Both plants were F1 or F2 progenies of 'Idared' × MR5. Colour picture at http://e-collection.library.ethz.ch/view/eth:6104

Malus × robusta 5

Malus × robusta (Carrière) Rehder is a hybrid of *M. baccata* and *M. prunifolia* (Gardner et al. 1980), both originating in South-Eastern Asia (Way et al. 1991). *Malus × robusta* 5 (MR5) was selected in the 1920ies from open pollinated *Malus × robusta (Watkins and Spangelo 1970).* Synonymous names are "*M. × robusta* Robusta 5", "*Malus microcarpa* Carrière var. *robusta* Carrière" and "*Pyrus baccata* L. var. *cerasifera* Regel" (Germplasm Resources Information Network (GRIN), http://www.ars-grin.gov). MR5 was described as "fire blight resistant, easy to propagate, winter hardy, outstanding anchorage, and non-brittle roots" (Watkins 1971), and it was widely used in rootstock breeding and apple research. For example MR5 was included in the Geneva apple rootstock breeding program and it is a parent from many Cornel-Geneva advanced rootstock selections (Norelli et al. 2003). The resistance towards FB is described differently, always depending to the strain of *E. amylovora* used for inoculation. The resistance level ranges from 100% resistance (Peil et al. 2007) to almost completely susceptible if inoculated with resistance breaking strains (Norelli and Aldwinckle 1986; Peil et al. 2011). Two strains of *E. amylovora* are known to break the resistance of MR5. (1) Ea266 (synonym E4001 A), isolated in Ontario (Canada) on a 'Rhode Island Greening Apple', broke the resistance with a shoot infection of 93% (Norelli and Aldwinckle 1986) and recently (2) Ea3049 (also known as COCPB265, Ea265 and E2002a, respectively), also isolated in Canada on apple cultivar 'Jonathan' (Lecomte et al. 1997), was reported to break the resistance with 97% shoot necrosis (Peil et al. 2011).

Erwinia amylovora

Bacteria of the genus *Erwinia* belong to the enterobacteria of which many are pathogenic to animals, humans (e.g. several *E. coli* strains, *Salmonella* species) and plants (e.g. *Pantoe* species). The genus *Erwinia* comprises besides the FB causing *E. amylovora* (Burr.) also *E. pyrifoliae*, a pathogen of Asian pear tree, *E. tasmaniensis*, an epiphytic bacterium (Geider et al. 2006), the nonpathogenic *E. billingiae* (Mergaert et al. 1999) and others. *Erwinia amylovora* affects about 200 species of the subfamily Spiroideae like the ornamental hosts *Crataegus* (hawthorn), *Cotoneaster*, *Pyracantha* (firethorn), *Malus*' crab-apples and *Photinia*, but also raspberry, blackberry and Japanese plum (*Prunus salicina*) (summarized by Kim and Beer 2000). *E. amylovora* overwinters in cankers and symptomless carriers from where it is primarily spread by rain and insects such as ants and flies to new infection sites (wounds and natural openings like nectarthodes of flowers) (Thomas and Ark 1934; Vanneste and Eden-Green 2000). From flowers it is mainly pollinator dispersed but the bacterial ooze (Figure 1.3A) coming out of infected tissues is also spread by rain and birds. *E. amylovora* can colonize the entire tree starting from one infected flower and kill the tree within a single growth season (Vanneste 2000). The bacteria migrate inside the host tissue and left behind tissue will necrotize rapidly (Figure 1.4) (Vanneste and Eden-Green 2000). Moreover *E. amylovora* migrate toward the rootstock causing disease with no or hardly recognizable symptoms (Thomson 2000) and therefore resistant rootstocks would also represent an advantage in apple production. Amylovoran (Bellemann and Geider 1992) and levan (Gross et al. 1992) are exopolysaccharides produced by pathogenic erwinias and giving them their species name.

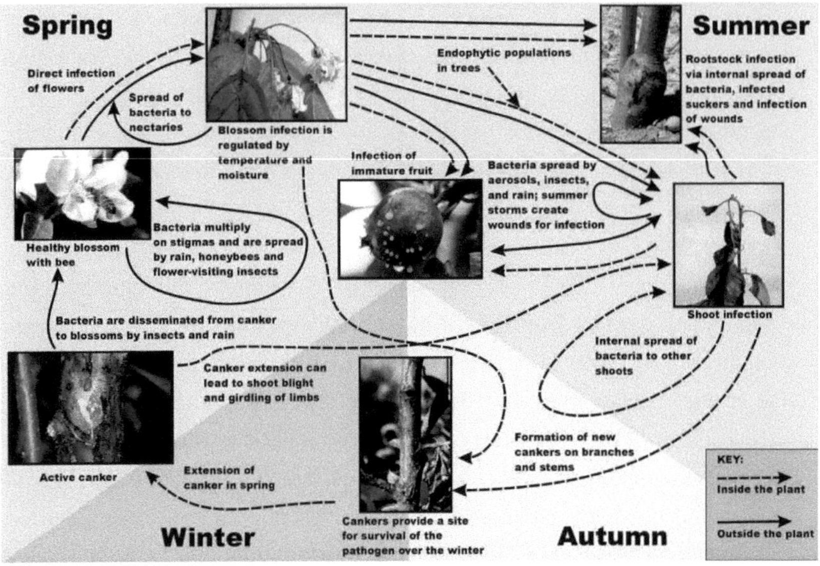

Figure 1.4 Disease cycle of fire blight during four seasons. Dashed lines represent movements of the causal agent, the bacterium *E. amylovora*, inside of the plant; solid lines represent the movements outside of plants. Disease cycle taken from Norelli et al. (2003).

The genome sequence of *E. amylovora* has been recently released (Smits et al. 2010) and may give additional information about putative pathomechanisms through the identification of genes coding for e.g. Avr gene products or other effectors. At GenBank (http://www.ncbi.nlm.nih.gov/genbank) five sequences of *Erwinia* were deposited: *E. amylovora* ATCC 49946 (Sebaihia et al. 2010) and CFBP1430 (Smits et al. 2010), *E. pyrifoliae* DSM 12163^T (Smits et al. 2010), *E. tasmaniensis* Et1/99 (Kube et al. 2008) and *E. billingiae* Eb661 (Kube et al. 2010). Smits and colleagues (2010) detected several secretion machineries: Type I, type II, type III and type VI. In *E. amylovora* several proteins are delivered through the T3SS but their recognition as effector is unknown: HrpJ, HrpQ, HrpA, HrpN, HrpW, HopPtoC, Eop2 (HokAK1Ea);

and several T3 effectors are described: AvrRpt2$_{EA}$, DspA, DspE, Eop1, Eop3 (HopX1Ea), EopB and OrfB (Kim and Beer 2000; Dean 2011; Khan et al. 2011; Bocsanczy et al. 2012). DspA/E (Gaudriault et al. 1997; Bogdanove et al. 1998), AvrRpt2$_{EA}$ (Zhao et al. 2006) and HopPtoC (Zhao et al. 2005) are known to be crucial pathogenicity factors of *E. amylovora*. Other proteins like the Eop1 and Eop3 are under discussion as pathogenicity factors (Nissinen et al. 2007; Bocsanczy et al. 2012). As already stated, AvrRpt2$_{EA}$, was demonstrated to elicit HR in *A. thaliana* (Zhao et al. 2006).

As described above the disease is controlled partially by pruning, eradication and copper application and completely with the use of antibiotics, whereas the latter selects strongly for antibiotic resistant *E. amylovora* strains. The bacterial resistance towards antibiotics is genetically implemented in the chromosome or located on plasmids harbored by the bacterium. The *R* genes can be spread vertically but also horizontally between bacterial classes by transformation and conjugation or phage mediated transduction (reviewed in Paget and Simonet 1994). If a DNA fragment coding e.g. for resistance is taken up by a competent bacterium, this gene may be integrated by flanking transposons or integrons into the genome (reviewed in Ochman et al. 2000). In *E. amylovora* antibiotic resistant strains were first detected in the 1970 in the USA and today commonly found in the western parts of the US (McManus et al. 2002). Streptomycin resistance is mostly manifested in mutations of the genomic *rpsL* gene which codes for the ribosomal protein S12 (Chiou and Jones 1995). Irrespective of their habitat in agricultural areas, resistant bacteria strains are a potential risk for human health. The self-transmissible plasmid pEa34 confers resistance against streptomycin to *E. amylovora*. The latter resistance is linked to a Tn3-type transposon which is able to integrate the resistance to other plasmids and chromosomes (McManus et al. 2002). In areas with occurrence of streptomycin resistant *E. amylovora* strains the disease is controlled with the less effective oxytetracyclin or a combination of streptomycin and oxytetracyclin (McManus et al. 2002).

The potential loss of antibiotic effectiveness against *E. amylovora* illustrates a dormant risk for apple producers. Therefore alternative measures should be developed. The breeding or genetic engineering of resistant fruit trees is the aim of many research programs which makes a detailed understanding of host resistance mechanisms mandatory as prerequisite.

New Technical Breeding and Engineering Strategies

Fire blight resistance is due to the high damaging potential of the disease one major breeding goal of many breeding programs. The pyramiding of resistance genes against one or multiple diseases is the advanced breeding goal (Kellerhals et al. 2009; Kellerhals et al. 2011). As described above the sources of FB resistance and other resistance traits are found often in wild and ornamental apple species and cultivars, respectively. Hence, after introgression of a resistance trait the main task for breeders remains to eliminate the unwanted 'wild' traits and the selection of commercially acceptable cultivars. This is a time-consuming task due to the disadvantage of the long generation cycle and cultivar self-incompatibility of apple trees. Additionally, the new developed and resistant apple plant will always be a new apple cultivar, which is hard to establish on the marked that is dominated by only a few cultivars. Adjacent to conventional breeding strategies different technical strategies are used to introgress resistance traits to salable cultivars with the aim to reduce laborious work, costs and breeding time or to bypass the development of a new cultivar. For those techniques the knowledge of a present resistance QTL is not always sufficient but the isolated gene is a premise. Two interesting and promising approaches will be highlighted in the following.
The first approach has been recently published by Flachowsky and co-workers (2011), in the PhD-thesis of Le Roux (2011) and by Le Roux et al. (2011). The authors developed transgenic apple lines containing an early

flowering gene from birch (*Betula pendula*), which allows the seedling plant to flower in less than one year instead of about four years as in non-transgenic apple plants. The transgenic plants have been crossed with several resistance donors resulting in six transgenic lines. Three lines carried *Rvi2*, *Rvi4* (scab resistance genes) and FB-F7 (FB resistance QTL of 'Fiesta' also present in and taken from cultivar 'Regia') and three lines Pl-1 and Pl-2 (powdery mildew resistance loci) as pyramided traits. Following the Mendelian laws of gene heritability, 50% of the obtained offspring will be transgenic and carries the birch gene and the other 50% will be non-transgenic apple plants. Therefore, after additional pseudo-backcrosses of the early flowering plants with high quality cultivars, the descendants without transgene but with the resistance loci and with high fruit quality can be chosen as resulting cultivar. This resulting cultivar would be a new one. This technique allows one generation per year instead of 3-8 years in conventional breeding.
The second approach is known as cisgenesis. Cisgenetic engineered plants are defined as plants harboring only genes from crossable donor plants that are under the regulation of their own promoter and terminator, and without any foreign genes such as marker genes used for transgene selection (Schouten et al. 2006). In contrary transgenic plants may harbor genes mixed of different organisms (e.g. Bt toxin of *Bacillus thuringiensis* in maize). The first development of a true cisgenic plant was recently published, describing the engineering of the apple cultivar 'Gala' with the *HcrVf2* gene of apple cv. 'Florina' (Vanblaere et al. 2011). Cisgenesis brings the advantage of the possibility to implement genes of interest in established cultivars and pyramid them in a shorter time than conventional breeding. The unpredictable locus of insertion of the target gene and its putative side effects remain as hot topics for current research. Under the existing European directive the cisgenic plants have the status of engineered transgenic plants which need to undergo costly tests before entering the public market (Schouten et al. 2006) and therefore this

engineering strategy may not be used for commercial purpose in near future (Gessler 2011).

Aim of the Thesis

Most apple growing regions have to struggle with fire blight. Due to the governmental limitations and resistance selection occurring with the application of antibiotics the disease management is still controversially discussed. The severe outbreak of fire blight in 2007 in Switzerland which could be only limited by the eradication of more than 100 hectares of commercial apple production area reminded to the high damage potential of FB and forced to release projects to find alternatives in fighting fire blight. Due to the lack of resistant apple cultivars, breeders and researchers investigated the ornamental and wild apple resources to identify usable resistance traits for breeding or genetic engineering. Markers tightly linked to major resistance QTLs as the one of MR5 can be used for marker-assisted selection (MAS) reducing laborious phenotypic evaluation. Especially if the trait of interest is introgessed from wild or ornamental relatives of *M.* × *domestica* and several pseudo-backcrosses are required to reach an acceptable fruit quality, MAS is an important tool for breeders. Furthermore only with MAS it is possible to pyramid functionally different resistances which is the current strategy to avoid or retard a break through by the pathogen.

Furthermore, cloned *R* gene(s) could be used in genetic engineering and characterization of the molecular interaction between *E. amylovora* and its host plants. Several major QTLs were detected with the two strongest ones in *Malus* × *domestica* cv. 'Evereste' on LG 12 and in *Malus* × *robusta* 5 on LG 3 (Durel et al. 2009; Peil et al. 2007). The resistance trait of 'Evereste' was studied recently and two putative resistance gene candidates were isolated (Parravicini et al. 2011).

The major objective of this thesis was the isolation and characterization of the strongest resistance QTL found on linkage group 3 of the wild apple species *Malus* × *robusta* 5, explaining up to 80% of the phenotypic variation of FB resistance in the cross of 'Idared' × MR5.

After the general introduction of the presented doctoral thesis, it is described in Chapter 2 the process and the results of fine mapping of the region of interest using a mapping population of more than 2,000 individuals which was followed by phenotyping of the informative recombinant individuals. Once the locus was narrowed to an appropriate size, the region containing putative resistance genes were isolated in a BAC library and sequenced. The resulting sequences were used to detect and clone resistance gene candidates from MR5 that were further characterized.

The third chapter summarizes the results of a comparison of the gene prediction on sequences derived from MR5 DNA carrying BAC with the sequenced transcriptome of unchallenged MR5. In addition the chapter describes the expression of the candidate resistance gene in MR5, selected progenies and susceptible cultivars using quantitative real time PCR.

In chapter four the functionality of the identified candidate *R* genes should be evaluated in a transgenic complementation assay. The assay comprises the *Agrobacterium*-mediated transformation of *in vitro* apple plants of the cultivar 'Gala' with different constructs containing the candidate gene(s) under the control of an over expression promoter and its own native promoter, respectively. After micro grafting of regenerated transgenic shoots future works will evaluate the status of fire blight resistance level to prove or to refuse the candidate genes.

In the last chapter, the General Conclusion and Discussion, the presented findings are summarized and reflected in a wider context.

References

Agroscope-Changins/Wädenswil (2012). "Fire blight." Retrieved January, 25th 2012, 2012, from http://www.agroscope.admin.ch/feuerbrand.

Ausubel, F. M. (2005). "Are innate immune signaling pathways in plants and animals conserved?" Nat. Immunol. 6(10): 973-979.

Axtell, M. J. and Staskawicz, B. J. (2003). "Initiation of RPS2-specified disease resistance in *Arabidopsis* is coupled to the AvrRpt2-directed elimination of RIN4." Cell 112(3): 369-377.

Bellemann, P. and Geider, K. (1992). "Localization of transposon insertions in pathogenicity mutants of *Erwinia amylovora* and their biochemical characterization." J. Gen. Microbiol. 138(5): 931-940.

Bent, A. F. and Mackey, D. (2007). "Elicitors, effectors, and *R* genes: The new paradigm and a lifetime supply of questions." Annu. Rev. Phytopathol. 45(1): 399-436.

Bocsanczy, A. M., Schneider, D. J., DeClerck, G. A., Cartinhour, S. and Beer, S. V. (2012). "HopX1 in *Erwinia amylovora* functions as an avirulence protein in apple and is regulated by HrpL." J. Bacteriol. 194(3): 553-560.

Bogdanove, A. J., Bauer, D. W. and Beer, S. V. (1998). "*Erwinia amylovora* secretes DspE, a pathogenicity factor and functional AvrE homolog, through the Hrp (type III secretion) pathway." The Journal of Bacteriology 180(8): 2244-2247.

Bonman, J. M., Khush, G. S. and Nelson, R. J. (1992). "Breeding rice for resistance to pests." Annu. Rev. Phytopathol. 30(1): 507-528.

Bravin, E. and Kilchenmann, A. (2010). "Ländervergleich der Apfelproduktion." Agrarforschung Schweiz 1(2): 52-59.

Campbell, C. S., Evans, R. C., Morgan, D. R., Dickinson, T. A. and Arsenault, M. P. (2007). "Phylogeny of subtribe Pyrinae (formerly the Maloideae, Rosaceae): Limited resolution of a complex evolutionary history." Plant Syst. Evol. 266(1): 119-145.

Caplan, J., Padmanabhan, M. and Dinesh-Kumar, S. P. (2008). "Plant NB-LRR immune receptors: From recognition to transcriptional reprogramming." Cell Host Microbe 3(3): 126-135.

Chiou, S. and Jones, A. L. (1995). "Molecular analysis of high-level streptomycin resistance in *Erwinia amylovora*." Phytopathology 85: 324-328.

Collier, S. M. and Moffett, P. (2009). "NB-LRRs work a bait and switch on pathogens." Trends Plant Sci. 14(10): 521-529.

Dangl, J. L. and Jones, J. D. G. (2001). "Plant pathogens and integrated defence responses to infection." Nature 411(6839): 826-833.

Dean, P. (2011). "Functional domains and motifs of bacterial type III effector proteins and their roles in infection." FEMS Microbiol. Rev. 35(6): 1100-1125.

DeYoung, B. J. and Innes, R. W. (2006). "Plant NBS-LRR proteins in pathogen sensing and host defense." Nat. Immunol. 7(12): 1243-1249.

Di Matteo, A., Federici, L., Mattei, B., Salvi, G., Johnson, K. A., Savino, C., De Lorenzo, G., Tsernoglou, D. and Cervone, F. (2003). "The crystal structure of polygalacturonase-inhibiting protein (PGIP), a leucine-rich repeat protein involved in plant defense." Proceedings of the National Academy of Sciences 100(17): 10124-10128.

Dodds, P. N., Lawrence, G. J., Catanzariti, A.-M., Teh, T., Wang, C.-I. A., Ayliffe, M. A., Kobe, B. and Ellis, J. G. (2006). "Direct protein interaction underlies gene-for-gene specificity and coevolution of the flax resistance genes and flax rust avirulence genes." Proceedings of the National Academy of Sciences 103(23): 8888-8893.

Durel, C. E., Denance, C. and Brisset, M. N. (2009). "Two distinct major QTL for resistance to fire blight co-localize on linkage group 12 in apple genotypes 'Evereste' and *Malus floribunda* clone 821." Genome 52(2): 139-147.

Ellis, J., Dodds, P. and Pryor, T. (2000). "Structure, function and evolution of plant disease resistance genes." Curr. Opin. Plant Biol. 3(4): 278-284.

FAO (2009). "Food and agricultural organization of the United Nations Statistical database." Retrieved Aug. 9th, 2011, from http://faostat.fao.org.

Flachowsky, H., Le Roux, P.-M., Peil, A., Patocchi, A., Richter, K. and Hanke, M.-V. (2011). "Application of a high-speed breeding technology to apple (*Malus* × *domestica*) based on transgenic early flowering plants and marker-assisted selection." New Phytol. 192(2): 364-377.

Flor, H. (1971). "Current status of gene-for-gene concept." Annu. Rev. Phytopathol. 9: 275.

Forsline, P. L., Aldwinckle, H. S., Dickson, E. E., Luby, J. J. and Hokanson, S. C. (2010). "Collection, maintenance, characterization, and utilization of wild apples of Central Asia". In "Hortic. Rev.", John Wiley & Sons, Inc. 10.1002/9780470650868.ch1: 1-61.

Gardner, R. G., Cummins, J. N. and Aldwinckle, H. S. (1980). "Fire blight resistance in the Geneva apple rootstock breeding program." J. Am. Soc. Hortic. Sci. 105(6): 907-912.

Gaudriault, S., Malandrin, L., Paulin, J. P. and Barny, M. A. (1997). "DspA, an essential pathogenicity factor of *Erwinia amylovora* showing homology with AvrE of *Pseudomonas syringae*, is secreted via the Hrp secretion pathway in a DspB-dependent way." Mol. Microbiol. 26(5): 1057-1069.
Geider, K., Auling, G., Du, Z., Jakovljevic, V., Jock, S. and Völksch, B. (2006). "*Erwinia tasmaniensis* sp. *nov*., a non-phytopathogenic bacterium from apple and pear trees." Int. J. Syst. Evol. Microbiol. 56(12): 2937-2943.
Gessler, C. (2011). "Cisgenic disease resistant apples: A product with benefits for the environment, producer and consumer." Outlooks on Pest Management 22(5): 216-219.
Gessler, C. and Pertot, I. (2011). "Vf scab resistance of *Malus*." Trees - Structure and Function 10.1007/s00468-011-0618-y: 1-14.
Gómez-Gómez, L. and Boller, T. (2000). "FLS2: An LRR receptor-like kinase involved in the perception of the bacterial elicitor flagellin in *Arabidopsis*." Mol. Cell 5(6): 1003-1011.
Gross, M., Geier, G., Rudolph, K. and Geider, K. (1992). "Levan and levansucrase synthesized by the fireblight pathogen *Erwinia amylovora*." Physiol. Mol. Plant Pathol. 40(6): 371-381.
Gruber, M., Söding, J. and Lupas, A. N. (2006). "Comparative analysis of coiled-coil prediction methods." J. Struct. Biol. 155(2): 140-145.
Hans, T.-C. (2003). "Fresh insights into processes of nonhost resistance." Curr. Opin. Plant Biol. 6(4): 351-357.
Hayes, C. S., Aoki, S. K. and Low, D. A. (2010). "Bacterial contact-dependent delivery systems." Annu. Rev. Genet. 44(1): 71-90.
Heath, M. C. (2000). "Hypersensitive response-related death." Plant Mol. Biol. 44(3): 321-334.
Heng, Z., Ling, G., Yuxin, Y. and Huairui, S. (2008). "Review of the Chinese apple industry." Acta Hort. (ISHS) 772.
Holliger, E., Vogelsanger, J., Schoch, B., Duffy, B., LUSSI, L. and Bünter, M. (2007). "Das Feuerbrandjahr 2007." Schweiz. Obst-Weinbau 24.
Ignatov, A. and Bodishevskaya, A. (2011). "*Malus*". In "Wild Crop Relatives: Genomic and Breeding Resources". C. Kole, Springer Berlin Heidelberg 10.1007/978-3-642-16057-8_3: 45-64.
Janick, J., Cummins, J. N., Browa, S. K. and Hemmat, M. (1996). "Apples". In "Fruit Breeding, Tree and Tropical Fruits". J. Janick and J. N. Moore. 1: 1-77.

Jia, Y., McAdams, S. A., Bryan, G. T., Hershey, H. P. and Valent, B. (2000). "Direct interaction of resistance gene and avirulence gene products confers rice blast resistance." EMBO J. 19(15): 4004-4014.

Jones, D. A. and Jones, J. D. G. (1997). "The role of leucine-rich repeat proteins in plant defences". In "Adv. Bot. Res.". I. C. T. J.H. Andrews and J. A. Callow, Academic Press. 24: 89-167.

Jones, J. D. and Dangl, J. L. (2006). "The plant immune system." Nature 444(7117): 323-329.

Juniper, B. E. and Mabberley, D. J. (2006). "The story of the apple". Portland, Or., Timber Press.

Kajava, A. V. and Kobe, B. (2002). "Assessment of the ability to model proteins with leucine-rich repeats in light of the latest structural information." Protein Sci. 11(5): 1082-1090.

Kellerhals, M., Franck, L., Baumgartner, I. O., Patocchi, A. and Frey, J. E. (2011). "Breeding for fire blight resistance in apple." Acta Hort. (ISHS) 896: 385-389.

Kellerhals, M., Székely, T., Sauer, C., Frey, J. and Patocchi, A. (2009). "Pyramiding scab resistances in apple breeding " Erwerbs-Obstbau 51(1): 21-28.

Kessler, A. and Baldwin, I. T. (2002). "Plant responses to insect herbivory: The emerging molecular analysis." Annu. Rev. Plant Biol. 53: 299-328.

Khan, M., Zhao, Y. F. and Korban, S. (2011). "Molecular mechanisms of pathogenesis and resistance to the bacterial pathogen *Erwinia amylovora*, causal agent of fire blight disease in Rosaceae." Plant Mol. Biol. Report. 10.1007/s11105-011-0334-1: 1-14.

Kim, J. F. and Beer, S. V. (2000). "*hrp* genes and harpins of *Erwinia amylovora*: A decade of discovery". In "Fire blight the disease and its causative agent, *Erwinia amylovora*". J. L. Vanneste. Wallingford, CABI 10.1079/9780851992945.0055: 141-161.

Kobe, B. and Deisenhofer, J. (1994). "The leucine-rich repeat: A versatile binding motif." Trends Biochem. Sci. 19(10): 415-421.

Kopp, E. B. and Medzhitov, R. (1999). "The Toll-receptor family and control of innate immunity." Curr. Opin. Immunol. 11(1): 13-18.

Korban, S. S. (1986). "Interspecific hybridization in *Malus*." Hortscience 21(1): 41-48.

Kube, M., Migdoll, A., Gehring, I., Heitmann, K., Mayer, Y., Kuhl, H., Knaust, F., Geider, K. and Reinhardt, R. (2010). "Genome comparison of the epiphytic bacteria *Erwinia billingiae* and *E. tasmaniensis* with the pear pathogen *E. pyrifoliae*." BMC Genomics 11(1): 393.

Kube, M., Migdoll, A. M., Müller, I., Kuhl, H., Beck, A., Reinhardt, R. and Geider, K. (2008). "The genome of *Erwinia tasmaniensis* strain Et1/99, a non-pathogenic bacterium in the genus *Erwinia*." Environ. Microbiol. 10(9): 2211-2222.

Le Roux, P.-M., Flachowsky, H., Hanke, M.-V., Gessler, C. and Patocchi, A. (2011). "Use of a transgenic early flowering approach in apple (*Malus* × *domestica* Borkh.) to introgress fire blight resistance from cultivar Evereste." Mol. Breed. 10.1007/s11032-011-9669-4: 1-18.

Le Roux, P. M. (2011). "Molecular breeding for fire blight resistance in apple (*Malus* spp.)". IBZ Plant Pathology, ETH, Zurich, Doctor of Science, Document number in Press, pages: in Press

Lecomte, P., Manceau, C., Paulin, J.-P. and Keck, M. (1997). "Identification by PCR analysis on plasmid pEA29 of isolates of *Erwinia amylovora* responsible of an outbreak in Central Europe." Eur. J. Plant Pathol. 103(1): 91-98.

Liebhard, R., Koller, B., Gianfranceschi, L. and Gessler, C. (2003). "Creating a saturated reference map for the apple (*Malus* × *domestica* Borkh.) genome." TAG Theoretical and Applied Genetics 106(8): 1497-1508.

Mackey, D., Holt Iii, B. F., Wiig, A. and Dangl, J. L. (2002). "RIN4 interacts with *Pseudomonas syringae* type III effector molecules and is required for RPM1-mediated resistance in *Arabidopsis*." Cell 108(6): 743-754.

Martin, G. B., Bogdanove, A. J. and Sessa, G. (2003). "Understanding the functions of plant disease resistance proteins." Annu. Rev. Plant Biol. 54: 23-61.

McDonald, B. A. and Linde, C. (2002). "Pathogen population genetics, evolutionary potential, and durable resistance." Annu. Rev. Phytopathol. 40(1): 349-379.

McGhee, G. C. and Sundin, G. W. (2010). "Evaluation of kasugamycin for fire blight management, effect on nontarget bacteria, and assessment of kasugamycin resistance potential in *Erwinia amylovora*." Phytopathology 101(2): 192-204.

McManus, P. S., Stockwell, V. O., Sundin, G. W. and Jones, A. L. (2002). "Antibiotic use in plant agriculture." Annu. Rev. Phytopathol. 40(1): 443-465.

McManus, P. S., Stockwell, V. O., Sundin, G. W. and Jones, A. L. (2002). "Antibiotic use in plant agriculture." Annu. Rev. Phytopathol. 40: 443-465.

Mergaert, J., Hauben, L., Cnockaert, M. C. and Swings, J. (1999). "Reclassification of non-pigmented *Erwinia herbicola* strains from trees as *Erwinia billingiae* sp. *nov*." Int. J. Syst. Bacteriol. 49(2): 377-383.

Meyers, B. C., Morgante, M. and Michelmore, R. W. (2002). "TIR-X and TIR-NBS proteins: Two new families related to disease resistance TIR-NBS-LRR proteins encoded in *Arabidopsis* and other plant genomes." Plant J. 32(1): 77-92.

Moffett, P. (2009). "Mechanisms of recognition in dominant *R* gene mediated resistance". In "Adv. Virus Res.". L. Gad and P. C. John, Academic Press. Volume 75: 1-33, 228-229.

N'Diaye, A., Van de Weg, W., Kodde, L., Koller, B., Dunemann, F., Thiermann, M., Tartarini, S., Gennari, F. and Durel, C. (2008). "Construction of an integrated consensus map of the apple genome based on four mapping populations." Tree Genetics & Genomes 4(4): 727-743.

Ngugi, H. K., Lehman, B. and Madden, L. V. (2011). "Multiple treatment meta-analysis of products evaluated for control of fire blight in the eastern United States." Phytopathology 101(5): 512-522.

Nissinen, R. M., Ytterberg, A. J., Bogdanove, A. J., Van Wijk, K. J. and Beer, S. V. (2007). "Analyses of the secretomes of *Erwinia amylovora* and selected *hrp* mutants reveal novel type III secreted proteins and an effect of HrpJ on extracellular harpin levels." Mol. Plant Pathol. 8(1): 55-67.

Nomura, K., Melotto, M. and He, S.-Y. (2005). "Suppression of host defense in compatible plant–*Pseudomonas syringae* interactions." Curr. Opin. Plant Biol. 8(4): 361-368.

Norelli, J. L. and Aldwinckle, H. S. (1986). "Differential susceptibility of *Malus* spp cultivars Robusta-5, Novole, and Ottawa-523 to *Erwinia amylovora*." Plant Dis. 70(11): 1017-1019.

Norelli, J. L., Holleran, H. T., Johnson, W. C., Robinson, T. L. and Aldwinckle, H. S. (2003). "Resistance of Geneva and other apple rootstocks to *Erwinia amylovora*." Plant Dis. 87(1): 26-32.

Norelli, J. L., Jones, A. L. and Aldwinckle, H. S. (2003). "Fire blight management in the twenty-first century - Using new technologies that enhance host resistance in apple." Plant Dis. 87(7): 756-765.

NZZ (2007). Über 50 Millionen Franken Schaden durch Feuerbrand. NZZ Online. Zurich.

Ochman, H., Lawrence, J. G. and Groisman, E. A. (2000). "Lateral gene transfer and the nature of bacterial innovation." Nature 405(6784): 299-304.

Paget, E. and Simonet, P. (1994). "On the track of natural transformation in soil." FEMS Microbiol. Ecol. 15(1-2): 109-117.

Parravicini, G. (2010). "Candidate genes for fire blight resistance in apple cultivar 'Evereste'". IBZ Plant Pathology, ETH, Zurich, Doctor of Sciences, Document number 19203, pages: 149

Parravicini, G., Gessler, C., Denance, C., Lasserre-Zuber, P., Vergne, E., Brisset, M. N., Patocchi, A., Durel, C. E. and Broggini, G. A. L. (2011). "Identification of serine/threonine kinase and nucleotide-binding site-leucine-rich repeat (NBS-LRR) genes in the fire blight resistance quantitative trait locus of apple cultivar 'Evereste'." Mol. Plant Pathol. 12(5): 493-505.

Pearce, G., Strydom, D., Johnson, S. and Ryan, C. A. (1991). "A polypeptide from tomato leaves induces wound-inducible proteinase inhibitor proteins." Science 253(5022): 895-897.

Peil, A., Bus, V. G. M., Geider, K., Richter, K., Flachowsky, H. and Hanke, M.-V. (2009). "Improvement of fire blight resistance in apple and pear." International Journal of Plant Breeding 3(1): 1-27.

Peil, A., Flachowsky, H., Hanke, M.-V., Richter, K. and Rode, J. (2011). "Inoculation of *Malus* × *robusta* 5 progeny with a strain breaking resistance to fire blight reveals a minor QTL on LG5." Acta Hort. (ISHS) 896: 357-362.

Peil, A., Garcia-Libreros, T., Richter, K., Trognitz, F. C., Trognitz, B., Hanke, M. V. and Flachowsky, H. (2007). "Strong evidence for a fire blight resistance gene of *Malus robusta* located on linkage group 3." Plant Breed. 126(5): 470-475.

Potter, D., Eriksson, T., Evans, R. C., Oh, S., Smedmark, J. E. E., Morgan, D. R., Kerr, M., Robertson, K. R., Arsenault, M., Dickinson, T. A. and Campbell, C. S. (2007). "Phylogeny and classification of Rosaceae." Plant Syst. Evol. 266(1): 5-43.

Ragan, W. H. (1905). "Nomenclature of the apple : A catalogue of the known varieties referred to in American publications from 1804 to 1904". Washington, [D.C.], U.S. G.P.O.

Rairdan, G. J., Collier, S. M., Sacco, M. A., Baldwin, T. T., Boettrich, T. and Moffett, P. (2008). "The coiled-coil and nucleotide binding domains of the potato Rx disease resistance protein function in pathogen recognition and signaling." The Plant Cell Online 20(3): 739-751.

Sacco, M. A., Mansoor, S. and Moffett, P. (2007). "A RanGAP protein physically interacts with the NB-LRR protein Rx, and is required for Rx-mediated viral resistance." The Plant Journal 52(1): 82-93.

Scheer, J. M. and Ryan, C. A. (2002). "The systemin receptor SR160 from Lycopersicon peruvianum is a member of the LRR receptor kinase family." Proceedings of the National Academy of Sciences 99(14): 9585-9590.

Schouten, H. J., Krens, F. A. and Jacobsen, E. (2006). "Cisgenic plants are similar to traditionally bred plants." EMBO Rep. 7(8): 750-753.

Sebaihia, M., Bocsanczy, A. M., Biehl, B. S., Quail, M. A., Perna, N. T., Glasner, J. D., DeClerck, G. A., Cartinhour, S., Schneider, D. J., Bentley, S. D., Parkhill, J. and Beer, S. V. (2010). "Complete genome sequence of the plant pathogen *Erwinia amylovora* strain ATCC 49946." The Journal of Bacteriology: JB.00022-00010.

Smits, T. H. M., Jaenicke, S., Rezzonico, F., Kamber, T., Goesmann, A., Frey, J. E. and Duffy, B. (2010). "Complete genome sequence of the fire blight pathogen *Erwinia pyrifoliae* DSM 12163(T) and comparative genomic insights into plant pathogenicity." BMC Genomics 11: -.

Smits, T. H. M., Rezzonico, F., Kamber, T., Blom, J., Goesmann, A., Frey, J. E. and Duffy, B. (2010). "Complete genome sequence of the fire blight pathogen *Erwinia amylovora* CFBP 1430 and comparison to other *Erwinia* spp." Mol. Plant. Microbe Interact. 23(4): 384-393.

Swiderski, M. R., Birker, D. and Jones, J. D. (2009). "The TIR domain of TIR-NB-LRR resistance proteins is a signaling domain involved in cell death induction." Mol. Plant. Microbe Interact. 22(2): 157-165.

Takken, F. L., Albrecht, M. and Tameling, W. I. (2006). "Resistance proteins: Molecular switches of plant defence." Curr. Opin. Plant Biol. 9(4): 383-390.

Takken, F. L. W. and Tameling, W. I. L. (2009). "To nibble at plant resistance proteins." Science 324(5928): 744-746.

Tameling, W. I. L., Vossen, J. H., Albrecht, M., Lengauer, T., Berden, J. A., Haring, M. A., Cornelissen, B. J. C. and Takken, F. L. W. (2006). "Mutations in the NB-ARC domain of I-2 that Impair ATP hydrolysis cause autoactivation." Plant Physiol. 140(4): 1233-1245.

Thomas, H. E. and Ark, P. A. (1934). Fire blight of pears and related plants. University of California Agricultural Experiment Station Bulletin 586. Berkeley, California.

Thomson, S. V. (2000). "Epidemiology of fire blight". In "Fire blight the disease and its causative agent, *Erwinia amylovora*". J. L. Vanneste. Wallingford, CABI 10.1079/9780851992945.0055: 9-36.

Van der Biezen, E. A. and Jones, J. D. (1998). "Plant disease-resistance proteins and the gene-for-gene concept." Trends Biochem. Sci. 23(12): 454-456.

Van der Hoorn, R. A. L. and Kamoun, S. (2008). "From guard to decoy: A new model for perception of plant pathogen effectors." Plant Cell 20(8): 2009-2017.

Van der Plank, J. (1963). "Plant disease: Epidemics and control". New York and London, Academic Press.

Van der Zwet, T. (2006). "Present worldwide distribution of fire blight and closely related diseases." Acta Hort. (ISHS) 704: 35-36.

Van Ooijen, G., Mayr, G., Kasiem, M. M. A., Albrecht, M., Cornelissen, B. J. C. and Takken, F. L. W. (2008). "Structure–function analysis of the NB-ARC domain of plant disease resistance proteins." J. Exp. Bot. 59(6): 1383-1397.

Van Ooijen, G., Van den Burg, H. A., Cornelissen, B. J. C. and Takken, F. L. W. (2007). "Structure and function of resistance proteins in *Solanaceous* plants." Annu. Rev. Phytopathol. 45(1): 43-72.

Vanblaere, T., Szankowski, I., Schaart, J., Schouten, H., Flachowsky, H., Broggini, G. A. L. and Gessler, C. (2011). "The development of a cisgenic apple plant." J. Biotechnol. 154(4): 304-311.

Vanneste, J. L. (2000). "What is fire blight? Who is *Erwinia amylovora*? How to control it?". In "Fire blight the disease and its causative agent, *Erwinia amylovora*". J. L. Vanneste. Wallingford, CABI 10.1079/9780851992945.0073: 1-6.

Vanneste, J. L. and Eden-Green, S. (2000). "Migration of *Erwinia amylovora* in host plant tissues". In "Fire blight the disease and its causative agent, *Erwinia amylovora*". J. L. Vanneste. Wallingford, CABI 10.1079/9780851992945.0073: 73-83.

Velasco, R., Zharkikh, A., Affourtit, J., Dhingra, A., Cestaro, A., Kalyanaraman, A., Fontana, P., Bhatnagar, S. K., Troggio, M., Pruss, D., Salvi, S., Pindo, M., Baldi, P., Castelletti, S., Cavaiuolo, M., Coppola, G., Costa, F., Cova, V., Dal Ri, A., Goremykin, V., Komjanc, M., Longhi, S., Magnago, P., Malacarne, G., Malnoy, M., Micheletti, D., Moretto, M., Perazzolli, M., Si-Ammour, A., Vezzulli, S., Zini, E., Eldredge, G., Fitzgerald, L. M., Gutin, N., Lanchbury, J., Macalma, T., Mitchell, J. T., Reid, J., Wardell, B., Kodira, C., Chen, Z., Desany, B.,

Niazi, F., Palmer, M., Koepke, T., Jiwan, D., Schaeffer, S., Krishnan, V., Wu, C., Chu, V. T., King, S. T., Vick, J., Tao, Q., Mraz, A., Stormo, A., Stormo, K., Bogden, R., Ederle, D., Stella, A., Vecchietti, A., Kater, M. M., Masiero, S., Lasserre, P., Lespinasse, Y., Allan, A. C., Bus, V., Chagne, D., Crowhurst, R. N., Gleave, A. P., Lavezzo, E., Fawcett, J. A., Proost, S., Rouze, P., Sterck, L., Toppo, S., Lazzari, B., Hellens, R. P., Durel, C.-E., Gutin, A., Bumgarner, R. E., Gardiner, S. E., Skolnick, M., Egholm, M., Van de Peer, Y., Salamini, F. and Viola, R. (2010). "The genome of the domesticated apple (*Malus* x *domestica* Borkh.)." Nat. Genet. 42(10): 833-839.

Watkins, R. (1971). "Apple rootstocks." Annual Report East Malling Research Station for 1970 92: 97-98.

Watkins, R. and Spangelo, L. P. S. (1970). "Components of genetic variance for plant survival and vigor of apple trees." Theor. Appl. Genet. 40: 195-203.

Way, R. D., Aldwinckle, H.S., Lamb, R.C., Rejman, A., Sansavini, S., Shen, T., Watkins, R., Westwood, M.N. and Yoshida, Y (1991). "Apples (*Malus*)". In "Genetic Resources of Temperate Fruit and Nut Crops". J. N. Moore and J. R. J. Ballington, Acta Hort. (ISHS). 290: 3-46.

Zhao, Y. F., Blumer, S. E. and Sundin, G. W. (2005). "Identification of *Erwinia amylovora* genes induced during infection of immature pear tissue." J. Bacteriol. 187(23): 8088-8103.

Zhao, Y. F., He, S. Y. and Sundin, G. W. (2006). "The *Erwinia amylovora avrRpt2(EA)* gene contributes to virulence on pear and *AvrRpt2(EA)* is recognized by *Arabidopsis RPS2* when expressed in *Pseudomonas syringae*." Mol. Plant. Microbe Interact. 19(6): 644-654.

Zipfel, C. (2008). "Pattern-recognition receptors in plant innate immunity." Curr. Opin. Immunol. 20(1): 10-16.

Chapter 2

A Candidate Gene for Fire Blight Resistance in *Malus* × *robusta* 5 is Coding for a CC-NBS-LRR

Abstract

Fire blight is the most important bacterial disease in apple (*Malus* × *domestica*) and pear (*Pyrus communis*) production. Today, the causal bacterium *Erwinia amylovora* is present in many apple and pear growing areas. We investigated the natural resistance of the wild apple *Malus* × *robusta* 5 against *E. amylovora*, previously mapped to linkage group 3. With a fine-mapping approach on a population of 2133 individuals followed by phenotyping of the recombinants from the region of interest, we developed flanking markers useful for marker-assisted selection. Open reading frames were predicted on the sequence of a BAC spanning the resistance locus. One open reading frame coded for a protein belonging to the NBS-LRR family. The *in silico* investigation of the structure of the candidate resistance gene against fire blight of *M.* × *robusta* 5, *FB_MR5*, led us hypothesize the presence of a coiled-coil region followed by an NBS and an LRR-like structure with the consensus 'LxxLx[IL]xxCxxLxxL'. The function of FB_MR5 was predicted in agreement with the decoy/guard model, that FB_MR5 monitors the transcribed RIN4_MR5, a homolog of RIN4 of *Arabidopsis thaliana*, that could interact with the previously described effector $AvrRpt2_{EA}$ of *E. amylovora*.

Chapter 2 was published as

Fahrentrapp J, Broggini GAL, Kellerhals M, Peil A, Richter K, Zini E, Gessler C, 2012, "A candidate gene for Fire Blight resistance in *Malus* × *robusta* 5 is coding for a CC-NBS-LRR", Tree Geneitc & Genomes, DOI 10.1007/s11295-012-0550-3.

Introduction

Annually, 71 Mio metric tons of apples are produced worldwide and cultivars of *Malus* × *domestica* cover an area of 4.8 Mio hectares of land (FAO, 2009). Fire blight (FB) is one of the most disastrous diseases in apple cultivation. The causative Gram-negative bacterium, *Erwinia amylovora*, affects apple (*M.* × *domestica*) and pear (*Pyrus communis*) orchards worldwide and causes important losses in production. 46 countries reported the presence of the bacterium (Van der Zwet, 2006). With an unpredictable periodicity FB causes severe losses of several million US Dollars and complete pomefruit orchards need to be eradicated. On the one hand, FB can be controlled by pruning and eradication, although pruning is often followed by recrudescence, whilst on the other, the application of different products reduce the risk of complete orchard loss. For example bio-control agents (e.g. *Bacillus subtilis*, *Pantoea agglomerans*), chemicals (e.g. Prohexadione-Ca, acibenzolar-S-methyl) and antibiotics (e.g. streptomycin, oxytetracyclin, kasugamycin) are used to protect the crops (Ngugi et al., 2011). The potential success of the control measure chosen depends upon the exact timing of the application. In Europe, the time of flowering is the period of highest infection risk due to the accessibility of *E. amylovora* to apple plants through the natural openings of the nectarthodes, leading to so-called 'blossom blight'. The most effective control method on short term is the application of antibiotics. But due to the problem of selection of resistant strains and potentially remaining residues in soil, honey and the fruit itself its use is strictly regulated in many countries. However, the selection of resistant strains of *E. amylovora* may result in loss of efficacy of the compound applied (McManus et al., 2002). Moreover, the potential risk increases that the resistance trait from *E. amylovora* will be transferred to human-pathogenic bacteria. The risk for humans is even greater if the drug class applied is also used in human medicine, as in the case of streptomycin and tetracyclins. In order to avoid the use of antibiotics current research is attempting to achieve natural host resistance to this disease in orchards, either by conventional breeding or by new breeding strategies, including genetic engineering.

Natural resistance to *E. amylovora* has been described in different wild *Malus* species and it has recently been used in conventional breeding programs (Kellerhals et al., 2009). In order to effectively introgress fire blight resistance into new selections the strength of the resistance, its heritability and its agronomic usefulness must be evaluated. Breeding programs mostly involve labor intensive and costly fire blight resistance phenotype assessments through inoculation assays. Knowing the position of the trait in the genome, the genetic marker that is tightly linked to a strong monogenetic resistance locus could be used to speed up the breeding process by reducing the number of genotypes that need to be phenotyped. If resistance genes from *Malus* would be available they could be used to develop potentially more marketable cisgenic plants that only harbor species-own genes (Schouten et al., 2006).

Quantitative trait loci analysis can be used to investigate the position of monogenic traits. In plant-pathogen interactions such resistance traits can be ascribed to resistance (*R*) genes. During host-pathogen interactions the host's R gene products may undergo a so called gene-for-gene interaction with pathogenic effectors. This interaction may be direct between an R gene product and a pathogen derived avirulence (Avr) protein (Flor, 1971) or indirect. The indirect interaction is explained with the 'Guard-model' which describes the interaction of the Avr protein with the 'guardee' of the host which is monitored by the R protein (Van der Biezen and Jones, 1998). The 'Decoy-model' was recently introduced in which the guardee is replaced by the 'decoy' but in contrary to the guardee, the decoy is an alternative substrate and not a fitness enhancing target of Avr (Van der Hoorn and Kamoun, 2008; Block and Alfano, 2011). The *R* genes can be classified into "seven major structural classes of *R* proteins" (Kruijt et al., 2005). Three of the *R* gene classes are known to be directed against bacteria: Kinases such as Pto, CC/TIR-NBS-LRR such as RPS2 and receptor-like kinases (RLK) such as Xa21 (Hammond-Kosack and Jones, 1997). The CNLs (CC-NBS-LRR) are further characterized by the N-terminal conserved non-TIR motifs (Bai et al., 2002), containing the penta peptide 'EDVID' often separating two coiled-coils (Rairdan et al., 2008). This region is followed by the P-loop region (synonyms: NBS and NB-ARC) containing 'Walker A' (kinase 1a) and 'Walker B' (kinase 2), RNBS-A, -B, -C and -D, GLPL and 'MHD' (DeYoung and Innes, 2006; Van Ooijen et al., 2008).

Quantitative trait loci (QTLs) analyses were performed to identify the genetic regions of putative resistance genes. In apple, major QTLs are found in the *M.* × *domestica* cultivars 'Florina' and 'Fiesta', *M. floribunda* 821, the ornamental cultivar 'Evereste' and the wild apple *M.* × *robusta* 5 (MR5) (reviewed in Khan et al., 2011). The strongest QTL reported is located on the LG 3 of MR5 (Peil et al., 2007). It explains up to 80% of the phenotypic variation between susceptibility (100% PLL, percent lesion length) and resistance (0% PLL) to the strain Ea222_JKI (Peil et al., 2007).

Previous studies evaluated the resistance of MR5 with different results. MR5 has been classified as a "cultivar[s] showing diverse degree of susceptibility according to the inoculated strain" (Paulin et al., 1993). Similar results were found by Norelli and Aldwinckle (1986), who reported that the PLLs varied from 0% PLL after Ea273 inoculation to 93% using E4001A (Ea266). The latter strain isolated in Canada was obviously able to overcome the resistance of MR5. Another strain from Canada, E2002A (synonyms Ea3049 and Ea265), applied to segregating population of 'Idared' × MR5, also broke the resistance of MR5 (Peil et al., 2011). The facts that the resistance was broken twice by single strains and the high explanation of the phenotypic variation of the QTL suggest the presence of a resistance type matching the gene-for-gene model.

Recently, Parravicini et al. (2011) described eight putative *R* genes of apple against *E. amylovora* in a small 79 kb region that accounted for the FB resistance of 'Evereste'. They hypothesized that the two most likely functional genes, a kinase- and a CNL-encoding gene, could function similarly to Pto-Prf of tomato, which confer resistance to *Pseudomonas syringae* pv. *tomato* (Martin et al., 1993; Parravicini et al., 2011). However, no complementation experiments to demonstrate the resistance function of these genes has been reported so far. In this paper we report the enrichment of the top *M.* × *robusta* 5 LG 3 with new molecular markers followed by a phenotypic evaluation of the FB resistance of 33 recombinants from 2133 seedlings of MR5 or of a resistant F1 of MR5 as the male parent crossed with susceptible mother plants. The locus responsible for resistance against FB was localized, spanned by MR5 DNA-harboring BACs and sequenced. The sequences were used for gene prediction. The transcribed mRNA was sequenced and the candidate gene was further analyzed *in silico* in order to model the mode of function.

Materials and Methods

Plant material

The susceptible *M.* × *domestica* cultivar 'Idared' was crossed with the fire blight resistant *M.* × *robusta* 5 (MR5) at the Julius-Kühn Institute (JKI, Dresden-Pillnitz, Germany). A total of 140 plants of this population, called 04208, were previously used for QTL analysis (Peil et al., 2007). A second and third cross of 'Idared' × MR5 was performed (called 09261 and 05211, respectively) to increase the total population size. Three resistant progenies (DA02_2,7, DA02_2,40 and DA02_1,27) of cross 04208 were crossed at Agroscope Changins-Wädenswil (ACW, Wädenswil Research Station, Switzerland) with the susceptible cultivar 'La Flamboyante' (sold as Mairac®) and the medium-level susceptible hybrid 'ACW11303' ((Arlet x Gloster) x Rewena), as well as with an uncharacterized mother plant. These crosses (0802-0804, 0805-0807, 0808; Table 2.1) were part of a breeding program towards resistance pyramiding at ACW (Switzerland). In total,

2137 plants were raised in the greenhouse, transplanted to the field and pots, respectively, and used for mapping (Table 2.1).

Table 2.1 Mapping population with a total number of 2137 individuals.

Population name	Crosses	n	Place of phenotyping experiment	Number of recombinants between markers FEM57 and FEM18
0802-0804	'ACW11303' × DA02_x	875	CH	18
0805-0807	'La Flamboyante' × DA02_x	511	CH	2
0808	XY × DA02_2,7	29	CH	0
04208	'Idared' × MR5	284	GER	6
09261	'Idared' × MR5	354	GER	5
AA01	'Idared' × MR5	84	GER	2

MR5 was the source of resistance in all crosses. 'ACW11303' was a medium-level susceptible accession from Agroscope Changins-Wädenswil that was used as a mother plant without a MR5 parentage. 'La Flamboyante' and 'Idared' were susceptible mother plants. DA02_x were resistant progenies of the population 04208 ('Idared' × MR5) phenotyped by Peil et al. (2007). The 'x' stands for one of three different pollen donor plants from the latter population with the numbers 2,7, 2,40 and 1,27, all progenies of MR5. XY is an unknown mother plant with unknown susceptibility. The resistance donors are underlined.

Marker enrichment in the region of interest

The region of interest (roi) was defined as the region with a high certainty of containing the postulated resistance gene(s). The roi was narrowed down by deep data mining of the phenotypic and genotypic data of the original dataset generated by Peil et al. (2007). We re-mapped the fire blight resistance trait as a single gene using the method of Durel et al. (2009) to transform the phenotypic data into binary data. Individuals with fewer than three repetitions during each phenotyping experiment or which showed a deviation in the average PLL between experimental repetitions of greater than 30% were rigorously excluded.

After DNA extraction (Frey et al., 2004), the SSR and SNP markers reported to map on the top of LG 3 or which were developed on the top of the LG 3 of the whole genome sequence (WGS) of 'Golden Delicious' (GD) (Velasco et al., 2010), were applied to a subset of 92 individuals in order to verify their location on top of LG 3 (Table 2.2). The amplification of SSRs was either performed using fluorescently labeled primers in conventional PCR or by using a method described by Schuelke (2000). For this method, the forward primer is prolonged by a nucleotide sequence similar to a universal primer, which is fluorescently labeled. As a universal primer we used either of the sequences FAM-gactgcgtaccaattcaaa or HEX-gactgcgtaccaattcacg. The SNP loci were amplified by conventional PCR and sequenced using the BigDye Terminator Kit 3.1 (Applied Biosystems, Carlsbad, California, U.S.). Amplified fragments of the SNPs and SSRs were analyzed on an ABI3730 or an ABI3310xl capillary sequencer (Applied Biosystems) using the software programs 'Sequencher v4.8' (Gene Codes Corporation, Ann Arbor, MI USA) and 'GeneMapper v4.0' (Applied Biosystems), respectively. Four SSR markers (FEM18, FEM57, Ch03e03, MdMYB12) selected on GD WGS to spread equally in the roi were used to screen the whole population of 2137 individuals in order to identify individuals showing recombination events in this interval. The other markers used for fine mapping were only applied to these recombinant individuals.

The genotypic data were processed and mapped using JoinMap3 (Van Ooijen and Voorrips, 2001) with a LOD (logarithm of odds) threshold of 10 and manually checked for consistency. For stability and a realistic estimate of distances in cM, the map was calculated using the true genotypic and phenotypic data of each recombinant, and amended with the true genotypic data for the four selected SSR markers from the rest of the whole population. All missing genotypic and phenotypic data were replaced by deduced data under the assumption of a parental (e.g. non-recombinant) genotype/phenotype.

Table 2.2 Molecular markers with primer sequences and allele sizes.

Marker	Source	Type	Forward Primer	Reverse Primer	Quality	Allele(s) in MR5	Allele(s) in Idared
Ch03g07	Liebhard et al. (2002)	SSR	AATAAGCATTCAAAGCAATCCG	TTTTTCCAAATCGAGTTTCGTT	perfect	125/145	125
Ch03e03	Liebhard et al. (2002)	SSR	GCACATTCTGCCTTATCTTGG	AAAACCCACAAATAGCGCC	imperfect	184/(204)/206	204/206
EH034548	Norelli et al. (2009)	SNP	GACCAATTTGGATCTTGTAACTCC	CAGCTACCAATGTAGCAGTTAATCC	N/A	N/A	N/A
FEM07	This work *	SSR	TGGCTGGTTACTCCTTCACC	AGAGGAGCACAGGGAAATCA	perfect	183/189	183/191
FEM09	This work *	SSR	TGGATGAATTTCAATTGGAGTAA	GGGGTAATAGAATCGCCCATA	perfect	203	201/203
FEM11	This work *	SSR	GGAGGGGAAGTGAGGAGAGT	AGGAGGGGAGGAGGATAATG	perfect	215	220
FEM12	This work *	SSR	CGGGTCGTGGACTAAGAAAA	GAAGCACGCATCATCTCCTT	perfect	228/236	236/244
FEM14	This work *	SSR	GTAGCAATTGGGTTGCGACT	TTTTCCTCAGGTTTCTCAGCA	imperfect	(179)/181/187	179/187
FEM18	This work *	SSR	AGAGCCACCAAAACCTGAGA	CGAAACGTCTCTTTCCTCCA	perfect	224/234	234/257
FEM19	This work *	SSR	ATTCGCTTTCGTGAGGAAGA	GGGGATGCTGCAAGTTTAAG	perfect	134/150	134/156
FEM47	This work *	SSR	CCAAATGTTGGGTTTCCACT	CTACACAGCTGGGGAGGAAG	imperfect	(194)/209/217	194/217
FEM53	This work *	SSR	CTCAGCGGCTCTGTCTTCTT	GCCTTCAAAATTCGATGCTC	perfect	166	165
FEM57	This work *	SSR	ACAGTCGGGTTTGAAGGAGA	CCACCCTGTTGAAGCAATC	perfect	183/189	189
MR-ARGH31-like	Baldi et al. (2004)	SSR	GATACAGTCGGGTTTGAAGGAG	TTAACTAGCATAGCCATCATGC	perfect	146/152	152
MdMYB12	Chagne et al. (2007)	SSR	CTCGGCAATCGGTAAAGCTA	TATGAACAGTGAAACCCTAACCCTA	perfect	150/178	150
rp16k15	This work	SNP	CACACACAAATTTTGCCTTATTC	TCAAATGTGTCATTTCTGCAAC	N/A	N/A	N/A
t16k15	This work	SNP	CTCCTCTTTCAACTCTTTGTCC	TCAAATTGAGGAGAGAGAACTAGTTTC	N/A	N/A	N/A

The alleles given in parentheses were co-amplified regions without polymorphisms. The alleles in coupling with the resistance trait of FB_MR5 are underlined. *, developed on GD WGS by IASMA, Instituto Agrario San Michele all'Adige, Italy.

Fire blight phenotyping

Four to twelve replicates (on average 9.5 plants grafted onto rootstock M9) of each recombinant individual were trained to a single actively growing shoot with at least 15 cm in length. After being transferred to the quarantine greenhouse each shoot was inoculated by cutting the two most juvenile and fully unfolded leaves with scissors dipped into a suspension of *E. amylovora* strain Ea222_JKI at 10^9 cfu/mL. The strain Ea222_JKI (previously named Ea222 (Richter and Fischer, 2000; Peil et al., 2007; Peil et al., 2008)) used in this study was isolated on *Cotoneaster* in the Czech Republic (Richter and Fischer, 2000) and is distinct to Ea222 isolated on *M.* × *domestica* cultivar '20 Ounce' (Norelli et al., 1988). Therefore the strain was renamed from Ea222 to Ea222_JKI (Gardiner et al., 2012). The length of the shoot and degree of necrosis were recorded 21 days after inoculation and transformed into the percentage of lesion length per shoot (PLL). The binary data transformation method described by Durel et al. (2009) was used. The experiments with recombinants of the populations 04208 and 09216 (all 'Idared' × MR5 progenies) were carried out at JKI (Quedlinburg, Germany). The greenhouse conditions were the same as described by Peil et al. (2007). MR5 and 'Idared' plants were used as positive and negative control plants, respectively. Recombinants from crosses with 'La Flamboyante'/'ACW11303' (Table 2.1) were inoculated at ACW (Wädenswil Research Station, Switzerland). At ACW the resistant parents (DA02_2,7, DA02_2,40 and DA02_1,27) and the susceptible *M.* × *domestica* cultivar 'Gala Galaxy' were used as controls. The greenhouse conditions during the whole experiment were 22 °C/18 °C (day/night) and 70% rel. humidity; at JKI no additional light was provided, at ACW 10 h 400 W light per day was provided during growth of the shoots to the experimental length. The raw data obtained in Germany and Switzerland were handled separately until they were transformed into binary values.

Isolation, sequencing and analysis of the region of interest

A BAC library harboring MR5 DNA in pCC1BAC (Invitrogen, Carlsbad, USA) with HindIII cloning sites was constructed by Amplicon Express, Pullman,USA. A total of 36,864 BAC clones with an average insert size of 145 kb, representing approximately 3.5 times the diploid apple genome (Velasco et al., 2010), were spotted onto nylon filters. Filter hybridization was performed using a radioactively labeled probe designed on the sequence flanking the SSR marker Ch03e03 (forward primer: ggcatttcttgctcttctgc, reverse primer: ttggcagctgcaacatagac). The procedure was described in detail by Patocchi et al. (1999), although New England Biolab's 'NEBlot®Kit' (Ipswich, USA) was used to label the probe. The BAC DNA was extracted and the RP and T7 extremities of the positive BACs were sequenced (Galli et al., 2010). After designing the primer using Primer3 (Rozen and Skaletsky, 1999) on these sequences, the amplicons were amplified on the population's parents and sequenced to detect polymorphic sites such as SNPs in order to verify the source area of the BACs on LG 3. The BAC insert sizes were estimated by both the summation of the size of the bands obtained by HindIII digestions and pulsed-field gel electrophoresis after NotI digest (Broggini et al., 2007). The BAC clones covering the resistance locus were sequenced by 454-pyrosequencing (GS FLX (Titanium), Roche, Basel, Switzerland) and assembled using Newbler (v2.3, module: gsAssembler, Roche) using a minimum overlap of 40 bp with 90% identity. Additional Sanger-sequencing for gap closure was applied. The sequences and contigs of 454-sequencing were assembled with Sequencher (v4.8, Gene Codes Corporation). The ORFs were predicted using FGENESH with algorithms for tomato, *A. thaliana* (dicot plants) and *V. vinifera* (Salamov and Solovyev, 2000). The predicted proteins were further analyzed with the BLASTx (Altschul et al., 1990) and MotifScan to predict their function and specific motifs (Pagni et al., 2007).

Analysis and expression of candidate genes

Candidates belonging to one of the three classes of resistance genes against bacteria were analyzed using consensus sequences of known motifs, i.e. in the case of CNL for the presence of the kinases Walker A and B, the pentapeptide EDVID or the NB-ARC 'switch' MHD (Walker et al., 1982; Van Ooijen et al., 2008). Coiled-coil regions were predicted using Coils/pCoils (Lupas et al., 1991; Lupas, 1996). Constitutionally transcribed total RNA was extracted from young, healthy (non-infected) apple leaves of MR5, 'Idared' and 'Golden Delicious' according to the instructions given by the manufacturer (Concert™ Plant RNA Reagent, Invitrogen). Potential DNA contamination was removed using the TURBO DNA-free™ Kit (Applied Biosystems). The quality and yield was determined and visualized with the NanoDrop spectrophotometer and agarose gel electrophoresis. The absence of DNA was proven by PCR with RNA as template and PCR with primers for elongation factor with cDNA as template (Szankowski et al., 2009). The RNA was reverse transcribed to cDNA using 'Maxima™ Reverse Transcriptase' (Fermentas, Waltham, USA) with oligo dT. For gene expression, primer pairs at the start (ATG-forward: ATGGGGGAGAGGCTTTTCTTGTGGCATTCCTCCAAG) and stop codons ($TGA_{(tomato)}$-reverse: TCAAATCATCTTCCAATCTATATCTATGTAAG and $TGA_{(dicot\&vitis)}$-reverse: TCACGGAAATCGACCACCACACCTGGCC) were designed to amplify the whole coding sequence. Amplified DNA was cloned into pTZ57/RP (Fermentas) and transformed into ccdB-survival cells (Invitrogen). Inserts of plasmids were sequenced by primer walking.

Results

Region of interest

By mining the phenotypic data, which were previously produced by Peil et al. (2007), 36 individuals of the total 140 were excluded from transformation into binary values due to low number of replicates, no experimental repetition or high average variability in PLL. The averages of phenotypic data from the remaining 104 individuals were transformed into binary data according to Durel et al. (2009), excluding a further 20 individuals around the median PLL. The fire blight resistance co-segregated with Ch03e03 in 83 individuals, and in one individual a recombination was observed, indicating the position of the fire blight resistance locus as being 1.2 cM distal to Ch03e03. Thus, we determined the region of interest (roi) as being the interval from a locus between Ch03g07/Ch03e03 and the close end of LG 3.

Marker enrichment in the region of interest

Ten markers (EH034548, FEM09, FEM11, FEM12, FEM14, FEM18, FEM53, FEM57, MR-ARGH31-like and MdMYB12) were validated on a subset (n=92) to map to the top of LG 3 (Figure 2.1 A). It was assuming in accordance to roi definition that the fire blight resistance locus is located around Ch03e03. Four markers (FEM57, FEM18, MdMYB12 and Ch03e03) which span the roi in regular intervals were used to screen a total of 2137 progenies. Four individuals showed different SSR-alleles to the estimated parental ones and therefore were classified as out-crossers and excluded from further analysis. 33 individuals showed a recombination in this interval. These 33 individuals were used for all subsequent studies. The validated markers plus three additional ones (FEM07, FEM19, FEM47) were applied to the recombinants and mapped (Figure 2.1 B). The top 1.5 cM of LG 3 was now defined by 14 markers, with a median distance between each other of 0.05 cM and the biggest inter marker distance of 0.4

cM between FEM53 and FEM07. The most distal marker of LG 3 was FEM57, which was found to lengthen the linkage group by 0.4 cM compared to the previously published map of MR5 (Peil et al., 2007).

Phenotyping and mapping of the resistance locus

Following *E. amylovora* inoculation the resistant parents MR5, DA02_1,27 and DA02_2,40 showed no lesions at all, whereas DA02_2,7, which was also carrying MR5 resistance, showed a PLL of 7%. The susceptible cultivar 'Idared' showed an average PLL of 52%, the parent 'ACW11303' a PLL of 14% and 'La Flamboyante' a PLL of 29%. The susceptible standard control 'Galaxy' showed a PLL of 77%. The tested recombinant genotypes of populations 04208, 09261 and 05211 (all progenies of 'Idared' × MR5) showed lesions between 0 to 77% of the shoots with a median of 1% and an average of 20% (Figure 2.2 A). The PLL of the recombinants of crosses 0802-0807 ranged from 0 to 25% (Figure 2.2 B), with a median of 12% and an average of 11%.

Single gene analysis with transformed phenotypic data indicated the position of the resistance locus between markers FEM14/FEM47 and EH034548 flanking an interval of 0.29 cM (Figure 2.1 C). This interval was defined by six recombinants, five of which were between the resistance locus and EH034548 (Figure 2.3). Linkage mapping placed the resistance locus at a map position of 0.58 cM away from FEM57. One individual (0804-174, Figure 2.3) was genetically susceptible but its phenotype was resistant (0% PLL).

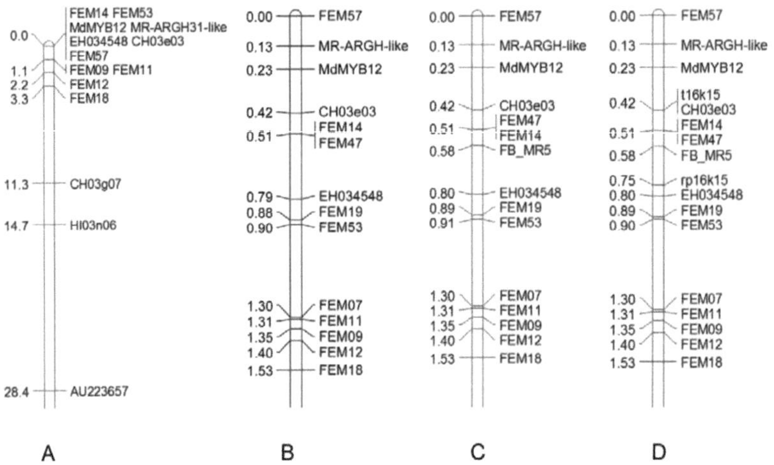

Figure 2.1 Marker enriched tops of linkage group 3 of *Malus* × *robusta* 5. A, mapped markers on top of linkage group 3 of a subset of 'Idared' × MR5 population, n=92. B, fine mapping of the top 1.5 cM of linkage group 3 on n=2133 individuals. C, FB_MR5 (fire blight resistance locus) mapped as a single gene after the transformation of phenotypic data into binary data; five recombinants were found between FB_MR5 and EH034548 and one between FB_MR5 and FEM47/FEM14. D, the genetic map as in C but SNPs were added which were developed on BAC end sequences (rp16k15 and t16k15).

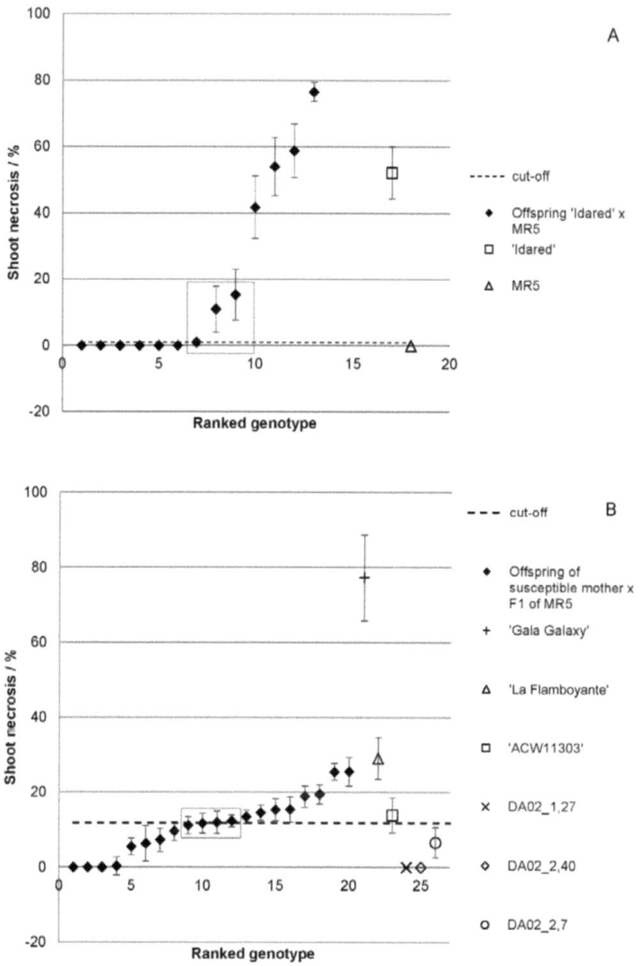

Figure 2.2 Necrosis values in % of shoot length of the phenotyped recombinants and controls. A, results of the experiments conducted at JKI (Germany). B, results obtained at ACW. All values are averages of 4-13 repetitions of each genotype. The filled diamonds show the recombinants, the different unfilled symbols show the controls. The data points, which were shown in rectangles, were excluded from the mapping process due to their position close to the cut-off value (median of average necrosis values of the genotypes tested). The cut-off values (A: 1% and B: 12%) between data transformation classes 'resistance' and 'susceptibility' are given as a black line. The bars display standard errors.

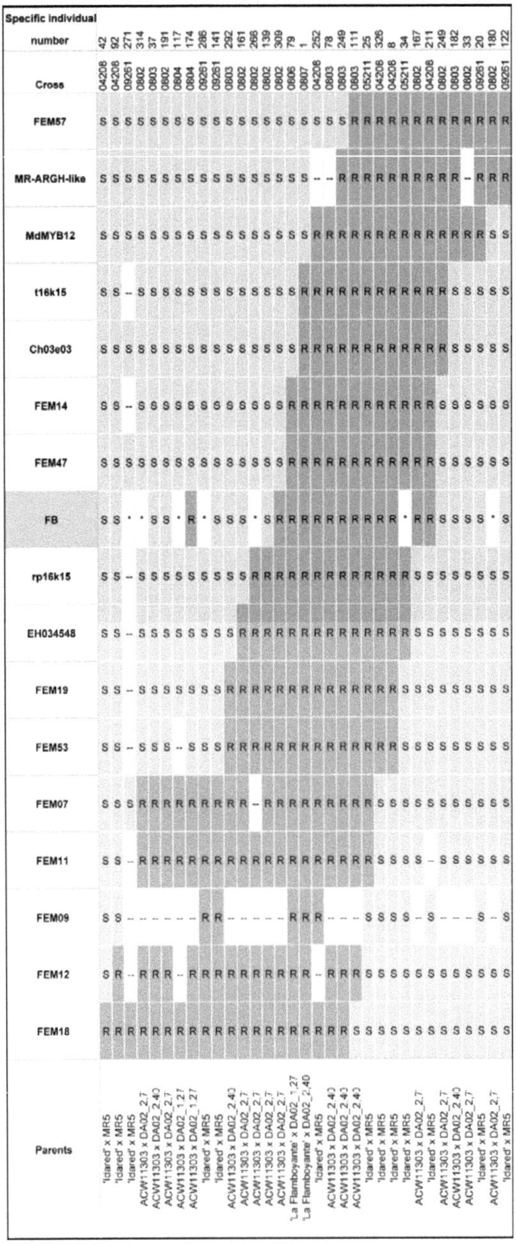

Figure 2.3 Genotypic and transformed phenotypic data of 33 progenies of *Malus × robusta* 5. The progenies were recombinant between FEM57 and FEM18. The R alleles (dark grey) were in coupling with phenotypic fire blight resistance (FB), S alleles in repulsion with FB (light gray); * indicates excluded phenotypic values (white) due to their position close to the cut-off value, -- stands for missing data (white). For gene mapping using the software JoinMap4, the missing data, which were not located at the border of a crossing over, were substituted by the allele of the missing marker which does not produce a crossing over towards his neighboring markers. The missing data of FEM09 in all crosses with 'ACW11303' were due to the lack of a polymorphism between 'ACW11303' and MR5 for this marker.

Chromosome landing and sequencing of the resistance region

A total of six clones were detected by hybridization of the genomic library with the probe derived from the sequence flanking Ch03e03. One of which, the BAC clone 16k15 (insert size 162 kb), carried flanking markers FEM14/FEM47, as found by means of PCR (data not shown). The extremities of this BAC clone were sequenced and SNP markers were developed on these sequences (t16k15 and rp16k15) and mapped to LG 3 (Figure 2.1 D). The marker rp16k15 was located one recombination closer to the FB resistance locus than EH034548, whereas the other extremity, t16k15, was located on the opposite side of the FB resistance locus. Therefore, the BAC 16k15 spanned the whole region of resistance locus. The marker rp16k15 was used as a new probe, which hybridized to another BAC named 72i24 (insert size 250 kb). This BAC clone carried the alleles in repulsion to the resistance locus of the flanking markers FEM14/FEM47 and rp16k15. Consequently, 72i24 spanned the same region of the homolog chromosome, which does not contribute to the resistance of MR5 against *E. amylovora*. Both BAC clones 16k15 and 72i24 were sequenced, resulting in a total of 162 kb in 4 contigs and 251kb in 22 contigs, respectively. The sequences of markers FEM14, FEM47, Ch03e03, t16k15 and rp16k15 were anchored to the four contigs of 16k15 (accession numbers HE805489-HE805492). The t16k15 and Ch03e03 sequences were anchored to one contig, whereas each of the remaining three contigs comprised one of the markers FEM14, FEM47 or rp16k15. The sequences of 72i24 were used as a database to detect the homolog sequences of the candidate genes.

Gene prediction

Three gene sets were predicted on the sequence of BAC 16k15 using different presets of FGENESH. The FGENESH algorithm for (1) *dicot* plants (based on *A. thaliana*) predicted 41 open reading frames (ORFs); (2) the algorithm for tomato predicted 47 ORFs and (3) the algorithm for *Vitis vinifera* predicted 25 ORFs. Genes with putative resistance function were identified with BLASTx (Table 2.2). A2.6 was found to be homolog to a putative serine/threonine kinase of *M.* × *domestica* (AEJ72559) but only with poor query coverage of 25% (68 of 272 amino acids). The sequences A4.11, A2.10 and V4.9 were found to share homologies with a TNL (TIR NBS-LRR) of *Arachis hypogaea*. Further A3.5 and T3.5 were homolog to a putative disease resistance RPP13-like protein 1-like of *V. vinifera*. All these BLASTx results could not be substantiated with MotifScan. Additional two predicted ORFs in each gene set (A2.2, A2.7, T2.2, T2.7, V2.1, V2.3) were similar to NB-ARC and LRR domains, which are common motifs of plant resistance genes. Only in A2.7, T2.7 and V2.3 the motifs (i.e.-ARC, Walker A and B, MHD) indicating functionality could be identified. The three predicted genes were located on the same sequence; all three shared the same start codon but showed two different splicing profiles, with the splicing and the stop codons predicted by the FGENESH*dicot* and *vitis* algorithms being the same. The mature transcript predicted by the FGENESH*dicot* and *vitis* algorithms contained one exon of 4089 bp followed by two shorter exons (69 and 153 bp, respectively). The FGENESH*tomato* algorithm predicted an unspliced transcript of 4167 bp with a different stop codon position. Therefore, the gene encoded in this transcript, which was predicted in two splicing variants, was the only putative candidate resistance gene.

Table 2.3 Hit with lowest E-value of BLASTx results of FGENESH predicted ORFs. Three sets of ORFs were predicted with FGENESH trained on sequences of *A. thaliana*, *V. vinifera* and tomato on a 162kb sequence (accession numbers HE805489-HE805492) derived from LG 3 of *Malus x robusta* 5

Predicted ORF[a]	Accession n°	Description	Species
A4.2	AAM22956	polycystin-1	*Canis lupus familiaris*
A1.6	ABU98601	beta-keratin 11	*Gekko gecko*
A4.8	ACF74323	unknown	*A. hypogaea*
T4.10	ACF74323	unknown	*A. hypogaea*
A2.2, A2.7, T2.2, FB_MR5, V2.1, V2.3	ADB66335	CNL protein	*Q. suber*
A2.6	AEJ72559	putative serine/threonine kinase	*M. x domestica*
A1.9	AEJ72571	hypothetical protein	*M. x domestica*
A4.11, T4.15, V4.9	AEL30371	TIR-NBS-LRR type disease resistance protein	*A. hypogaea*
T4.11, A2.10, T2.10	CAN63562	hypothetical protein VITISV_037178	*V. vinifera*
A4.3, T4.5, V4.2	CAN81355	hypothetical protein VITISV_039158	*V. vinifera*
A1.4, A2.13, A3.1, T2.13, V2.5, V3.1	CBI16489	unnamed protein product	*V. vinifera*
T2.1	CBI20229	unnamed protein product	*V. vinifera*
A2.8, T2.8, V2.4	CBI20561	unnamed protein product	*V. vinifera*
A1.2, T1.2, V1.2	CBL94163	putative RNA-directed DNA polymerase (Reverse transcriptase)	*M. x domestica*
A1.8, T4.14	CBL94164	predicted protein	*M. x domestica*
A4.9, V4.7	CBL94165	putative reverse transcriptase family member	*M. x domestica*
A4.10, T1.8, T1.9, T4.12, T4.13, V4.8	CBL94184	putative COBL7 (COBRA-LIKE 7)	*M. x domestica*

T4.2	CBY23069	unnamed protein product	*Oikopleura dioica*
T4.8	E47759	retrovirus-related reverse transcriptase homolog - soybean retrotransposon copia-like (fragment)	*Glycine max*
A2.1	EEH09863	conserved hypothetical protein	*Ajellomyces capsulatus* G186AR
V3.2	EGB04120	hypothetical protein AURANDRAFT_67470	*Aureococcus anophagefferens*
A2.9, A2.9, T4.4, T2.9, T3.6	not available	not available	
A1.5, A1.5, V1.5	NP_001172930	Os02g0330601	*Oryza sativa Japonica Group*
A4.6	XP_001604683	predicted: hypothetical protein LOC100121098	*Nasonia vitripennis*
A1.7, T1.7, T2.12, T3.2, V1.6	XP_002283424	predicted: uncharacterized protein LOC100246694	*V. vinifera*
A4.1, A4.5, A2.3, T4.1, T4.7, T2.3, V4.4, V2.2	XP_002283435	predicted: GPI mannosyltransferase 2	*V. vinifera*
A4.7, T4.9, V4.5	XP_002283442	predicted: non-specific lipid-transfer protein-like protein At2g13820-like	*V. vinifera*
A4.13, V4.11	XP_002283484	predicted: F-box/kelch-repeat protein At1g55270	*V. vinifera*
T1.4, V1.4	XP_002297647	predicted protein	*Populus trichocarpa*
T4.3	XP_002297779	predicted protein	*P. trichocarpa*
A1.1, A1.3, T1.1, T1.3, V1.1, V1.3	XP_002299130	predicted protein	*P. trichocarpa*
A4.4, A2.11, A2.14, A3.2, A3.4, T4.6, T2.11, T3.3, V4.3, V2.6	XP_002299228	predicted protein	*P. trichocarpa*
V4.1	XP_002304011	predicted protein	*P. trichocarpa*
A2.4, T2.4	XP_002327499	predicted protein	*P. trichocarpa*
T3.1	XP_002328020	predicted protein	*P. trichocarpa*
T2.14	XP_002328021	predicted protein	*P. trichocarpa*
A4.12, T4.17, V4.10	XP_002527054	conserved hypothetical protein	*Ricinus communis*

T4.18	XP_002527061	Protein AFR, putative	*R. communis*
A3.3	XP_002890343	predicted protein	*A. lyrata* subsp. *lyrata*
A2.5, T2.5	XP_003177448	hypothetical protein MGYG_08936	*Arthroderma gypseum* CBS 118893
T4.16	XP_003520891	predicted: tyrosine-sulfated glycopeptide receptor 1-like	*G. max*
V4.6[+]	XP_003524945	predicted: uncharacterized protein LOC100780285	*G. max*
T1.6	XP_003526328	predicted: uncharacterized GPI-anchored protein At1g51900-like	*G. max*
T3.4	XP_003562097	predicted: xylosyltransferase 1-like	*Brachypodium distachyon*
T2.6	XP_003598202	60S ribosomal protein L23	*Medicago truncatula*
A3.5, T3.5	XP_003633380	predicted: putative disease resistance RPP13-like protein 1-like	*V. vinifera*

[a]ORFs name starting with A, T and V were predicted with FGENESH trained either on *Arabidopsis*, Tomato or *Vitis*.

Expression of the putative candidate gene

The transcript of the putative resistance gene was amplified with primers on the ATG codon as well as two different primers designed on the two differently predicted stop codons. Amplicons were obtained using both primer pairs (ATG-forward, $TGA_{(tomato)}$-reverse, $TGA_{(dicot\&vitis)}$-reverse) and were cloned into the pTZ57/RP vector for sequencing, showing that the introns predicted by FGENESH*dicot* and *vitis* parameters were not predicted correctly. The full ORF of the transcript T2.7 was sequenced and called *FB_MR5* (accession number: CCH50986.1), being the only candidate resistance gene of *M.* × *robusta* 5 against *E. amylovora.* A homolog nucleotide sequence for this gene could not be amplified from BAC 72i24 nor from genomic DNA of the susceptible cultivars 'Idared' and 'Golden Delicious'.

Structure analysis of FB_MR5

Analysis with MotifScan and Coils/pCoils of FB_MR5 predicted N-terminal two coiled-coil regions (amino acid (aa) positions of around 60-75 and 120-150) separated by an EDVID-like motif (aa position 80) followed by a predicted NB_ARC (start at aa 172). The latter domain comprised the following motifs: Walker A and Walker B, RNBS-B, GLPL and MHD (Figure 2.4 A). At the C-terminal region behind the NB-ARC three LRRs were identified. The imperfect LRRs fulfilled the consensus LxxLxxLxLxxT and LxxLxxLxLxx(T/C)xxLxxIPxx, respectively, but lacked one 'Lxx' repeat at the start of the motif (Jones and Jones, 1997). The minimal consensus of LRRs 'LxxLxL' (Kajava and Kobe, 2002) increased the number of putative LRRs to eight. All eight LRRs were located behind MHD. In a *de novo* search for repetitive motifs using the software MEME (Bailey and Elkan, 1994), we identified the consensus 'LxSL[EKR]ELxIx[GD]CxSL' (Figure 2.4 C) in the amino acids behind NB-ARC (aa 732-1387). This consensus was refined to an LRR-like motif

'LxxLx[IL]xxCxxLxxL', of which a perfect version was present three times and an imperfect version 20 times (Figure 2.4 B). A hydropathy plot (Figure 2.5) resulted in values between -2 and +1.5, therefore no indications were found for membrane-spanning elements (Kyte and Doolittle, 1982). A protein BLAST analysis against NCBI's reference proteins found highest similarity (49%) of FB_MR5 to the predicted CNL of *Populus trichocarpa* (GenBank accession: XP_002328224). Protein BLAST against proteins predicted on GD genome sequence resulted in homology to three predicted peptide sequences on LG 3 of GD with highest similarity to MDP0000269188 of 62% in ClustalW pairwise alignment. MDP0000269188 comprises a NB-ARC domain lacking the motifs Walker B and MHD, and Walker A is present only in imperfect version. BLASTn of the candidate against apple ESTs identified GO532686 isolated from bark of 'Royal Gala' covering 15% of the query with 93% identity.

Figure 2.4 The structure and motifs of FB_MR5. FB_MR5 was identified as CNL which was subdivided in A: coiled-coil (CC) region and NB-ARC domain as well as B: LRR-like motifs. A, the CC regions and conserved motifs are underlined, NB-ARC is highlighted in gray. B, amino acids of the LRR-like motif are highlighted: L, blue; I, yellow; C, green; gray, residues that did not follow the consensus 'LxxLx[IL]xxCxxLxxL'. Classically known (Jones and Jones 1997) cytoplasmatic LRRs are underlined. C, motif logo created by *de novo* motif finder 'MEME' with the general consensus LxxLx[IL]xxCxxL. Amino acid '1' (arrow) corresponds to the first letter in each line of Figure 2.4 B. The colors of the letters indicate equal characteristics of amino acids; their height indicates the number of occurrences at the specific position. Colour picture at http://e-collection.library.ethz.ch/view/eth:6104

A

```
                                                                                  CC1
MGGEAFLVAFLQVLVDKLAHREVFKYFGLVKGVDQKLKKWSATLSAIGAVLNDAEERQLTAKNNTLKLWLEDLRDLAFDV
EDVID                                                   CC2
EDVLDKYATKMLKRQIQHAHSRTTSKLWNSIPDGVFNFNMNSEIQKISERLQEISEQKDQLNLKIDTGALTTRARRNISP
           NB-ARC                                      Walker A
SSSQPDGPVICRDEDKRKIVELLSKQEHRTVNFDVVAIVGMAGVGKTTLAGQVLNDMVATQTFQPAVWACVSDDFNLERV
                                                  Walker B                            RNBS-B
TKQILESITSRQCTTEDYNKVQDYLHKELAGKKFLIVLDDVWKTCSYGEWMKLQSPFRDGAQGSKIIVTTRDTDVSKMMG
                                                                GLPL
AATLVHNLEPMESSVCLQVFEQHAFLNSNDDKPPNYELLKEKIAAKCRGLPLAARTLGGVLLRKDTYEWEDILNNKLWSL

SNEHDILPVLRLTYFYLPSHLKRCFAYCSILPNDYEFEEKQMILLWMAEGFILPRPEDKKQIEDLGADYFRDLVSRSLFQ
          MHD
KSTKCISKYVMHDLIGDLARWAAGEICFRLEDKQNDDGEQLRCFPKARHSSYIRGLSDGVKRFEVFSELKYLRTFLPLRK
                                                         LRR1                   LRR2
DSFWNYLSRQVAFDLLPKLQYLRVLSFNCYKITELPDSIGDLRYLRYLDLSYTDITSLPKSTSTLYNLQTLILEGCSKLK

ALPIDMSNLVNLRHLNNSNVSLLEDMPPQLGRLVNLQSLTKFVVSGGGGGDRSGIRELEFLMHLRGTLCISRLENVTDVE

DAQRANLNCKER
```

B

```
LDSLVLEWSHSSDTRETESAVLDMLQPHTK
LKELTIKSYAGKEFSSWVGVPLFSNMVLVRLEECNNCLSLPPLGKLPH
LKELYIRGMNAVESVGAEFYGECSLPFPLLETLEFVDMQHWKVWLPFQTDHRGSVFPC
LKTLLVRKCSKLEGKLPENLDS
LASLEIVKCEELLVSIANYKQ
LRQLNIDGCKGVVHTAAKVEFEL
LESLYLSNISELTSLQTGELCRNGLNM
VRDLKINGCEELTSSLKNEAILLQQLIS
LGRLEIEDNSLLVEELGKEADELLQLQILGCK
LEFLKLKKCKNLLKLPEGLNQLSS
LQELRIHECSSLVSFPDVGLPPS
LKDIEITECHSLIYFAKSQIPQN
LRRIQIRDCRSLRSLVDNEAVGSCSSSSHNC
LEYLNIERCQSLTLLSLSDQLVRA
LRELDIYDCEQLEFLAPDGLFCNNTNYF
LENFRIRRCQNLKSLPRLSGGIRGSN
LREIRITDCDRLEALPEDMHNFNS
LEKLIIDYREGLTCSFPANLTS
LMIWKVKSCKSLWELEWGLHRLTS
LRYLWIGGEDPDMVSFPPDMVRMETLLPKS
LTELSIGGFPNLKKLSSKGFQFLTS
LESLELWDCPKLASIPKEGLPLS
LTELCIYGCPVLKERCQPGKGRYWHKISHIPYIDIDWKMI
```

C

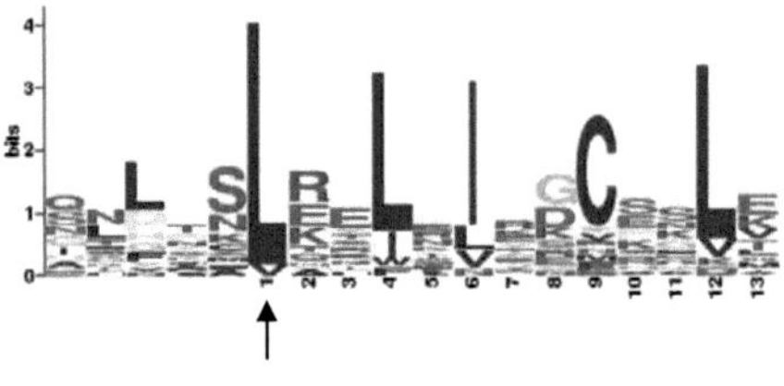

Figure 2.4

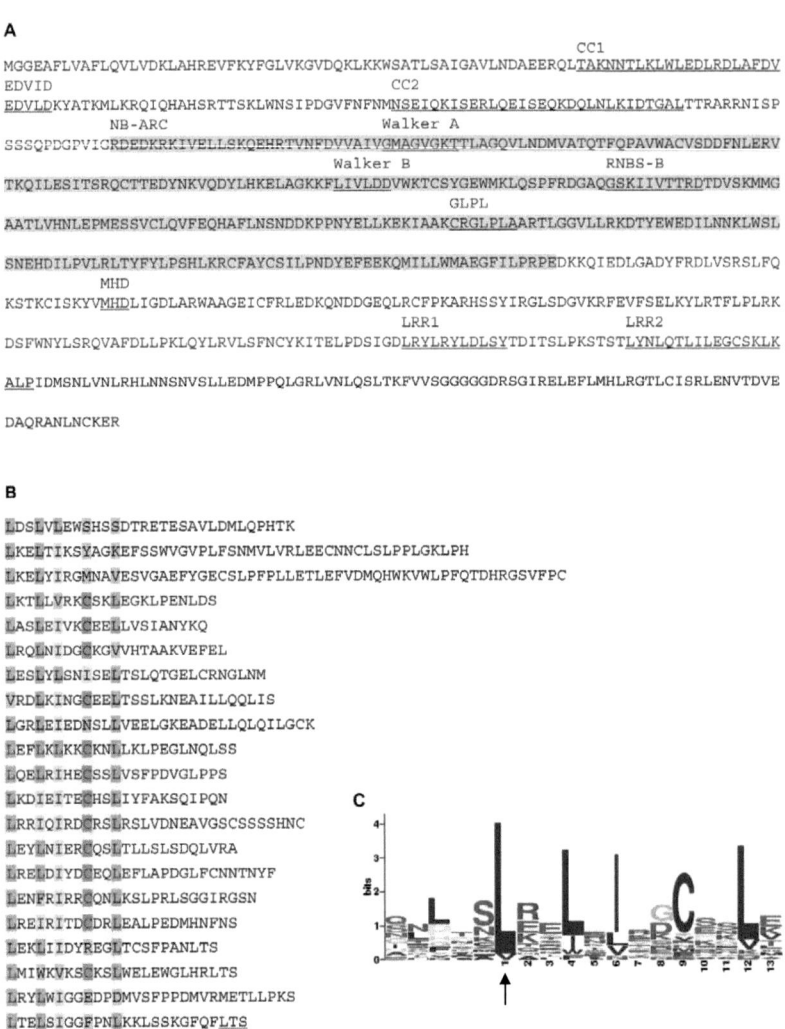

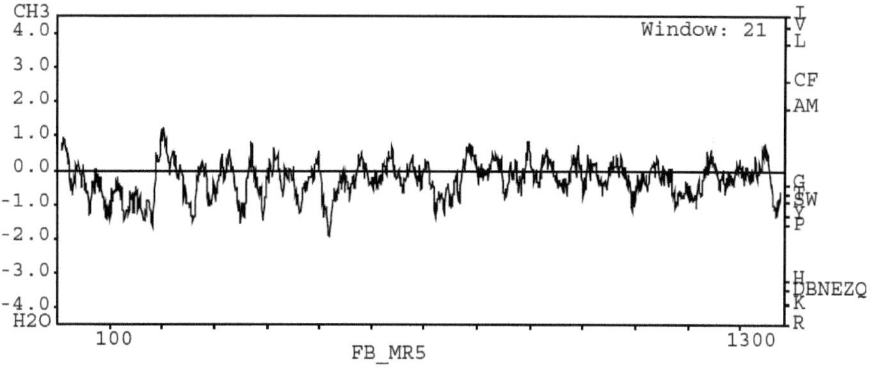

Figure 2.5 Hydropathy plot (Kyte and Doolittle 1982) of FB_MR5 with a window size of 21 aa. Values above 1.6 indicate a membrane-spanning protein domain which could not be identified for FB_MR5.

Discussion

Firstly, the original data of Peil et al. (2007) were investigated and used to precisely map the FB resistance trait as a single gene using the method published by Durel et al. (2009), which accounts for a certain degree of biological variance in these fire blight inoculation experiments. Then, the mapping population was increased to 2133 individuals, which allowed a mapping resolution of 0.05 cM. A total of 33 individuals were identified showing recombination in the region of interest on top of LG 3, where the resistance resides, and the resistance of these recombinants towards FB was assessed. This was done because only such individuals are informative in locating the exact map position of the resistance. Eventually, a genomic library of *M.* × *robusta* 5 was used to isolate the region of resistance, which was then sequenced and used to identify a putative resistance gene named *FB_MR5* belonging to the gene family encoding CNLs. Transcripts of this gene were reverse transcribed into cDNA, cloned into a vector and then sequenced to identify the correct splicing profile of this gene.

Fire blight remains the most feared disease in apple production as its appearance is highly erratic and the damage can potentially endanger every orchard. Classical breeding is currently introgressing resistance from sources such as *M.* × *robusta* 5 or *M.* × *domestica* cv 'Evereste'. As the selection of fire blight-resistant offspring by phenotypic reactions upon inoculation is labor intensive and costly DNA markers linked to the resistance trait represent an alternative selection tool for marker-assisted selection. Moreover, in order to evaluate the usefulness and durability of such resistance trait, the mechanisms of the plant-pathogen interaction need to be understood by studying the proteins involved in pathogen recognition, as well as the defense cascade induced. Finally, once a FB resistance gene has been identified and its functionality has been proven, it can be introduced into a commercial cultivar via genetic modification aiming to produce a cisgenic FB resistant apple cultivar where selectable marker genes are absent and only species-own genes are present with their own regulatory sequences from a crossable donor (Gessler, 2011). Cisgenesis has the additional benefits of maintaining all of the qualities that make a particular cultivar popular, whereas in classical apple breeding a new cultivar is created with its own unique qualities (Schouten et al., 2006). Recently, the first cisgenic apple cultivar carrying the *HcrVf2* scab resistance gene was reported (Vanblaere et al., 2011), and such a GM cultivar could be further improved by adding a fire blight resistance trait. Thus, for these reasons we attempted the positional cloning of the resistance against fire blight present in *M.* × *robusta* 5 (Peil et al., 2007). This resistance can be attributed to a single locus being a QTL explaining 80% of the phenotypic variation, and putatively it undergoes a gene-for-gene relationship as this resistance was broken twice by two different Canadian *E. amylovora* strains (Norelli and Aldwinckle, 1986; Peil et al., 2011). The resistance may still be useful as single major *R* gene in Asia,

Europe and other countries, where such strains have not yet been reported, or as pyramided FB resistance also in North America.

Mapping of the fire blight resistance locus

Re-analysis of the phenotypic and genotypic data produced by Peil et al. (2007) eliminated seven out of eleven individuals, which showed a disagreement >30% of the averaged PLLs between experimental repetitions. That confirmed that this method was useful and should be used in subsequent studies. The data of the remaining four individuals did not cause any double crossing overs. When the phenotypic data were mapped only one recombinant individual indicated a position of the FB locus distal to Ch03e03 (equal to 1.2 cM), which is in contrast with the previously mapped distance of 9 cM (Peil et al., 2007). Due to the remaining disagreement of the re-mapped FB locus regarding the maximum of the QTL (about 5 cM), the reduced dataset was used for a re-QTL analysis. This resulted in a LOD curve (data not shown) that increased towards the LG end, similar to that demonstrated for 'Evereste' (Durel et al., 2009) missing a maximum peak. This was probably caused by the lack of markers distal to Ch03e03 (MR5) and Hi23d11y_E ('Evereste'), respectively. After adding two new markers to LG 12 in 'Evereste' (M45TA_403c_E and M35TA_256s_E), 2.7 and 3.7 cM distal to Hi23d11y_E, respectively, the maximum of the LOD curve corresponded to the position resulting of the single gene analysis of FB_E (Parravicini, 2010). In order to solve the discrepancy in MR5 between the QTL curve maximum and the FB locus, the mapping population was increased to 2133 individuals and the roi was enriched with 15 new markers. As phenotyping for resistance/susceptibility against fire blight for a large number of progenies is cumbersome and costly, and since only individuals showing a recombination in the region of interest are informative in determining the map position of the resistance locus, we only considered these individuals with a crossing over in the roi that clearly encompassed the locus of the resistance trait inherited from

MR5. New QTL mapping of this data was not feasible because only the recombinants were phenotyped. The population used to map the resistance of MR5 was obtained by combining different populations resulting from the crossing of resistant parents (either MR5 or one of its resistant F1) with highly susceptible and medium-level susceptible parents. The susceptibility levels of the mother plants (average PLL of 'Idared' 52%, 'La Flamboyante' 29%, 'ACW11303' 14%) were clearly distinguishable from the resistance of MR5 and the resistant progenies of MR5 (average PLL 1.75%) used as male plants. Separate single gene analysis of the recombinants produced by each mother plant showed the same position of the fire blight resistance locus and therefore acknowledged that the different populations could be pooled (data not shown).

The application of the method suggested by Durel et al. (2009) to transform the phenotypic data into binary data eliminated three and four individuals, respectively, from each inoculation experiment (Figure 2.2 A and B). The individual 0804.174 which showed a phenotype incongruent to its phenotype was not excluded by this data treatment. Taking into account their standard error of the mean (SEM) as an indicator of the putative switch of ranked classes after experimental repetition if the SEM crosses the cut-off value, one additional individual on both sides of the four excluded individuals of ACW subpopulation should be eliminated. Also, 'ACW11303' had a large SEM ranging across the threshold towards the resistant category, indicating the lower susceptibility level of this parent. The resistance contributed by 'ACW11303' might be a quantitative resistance what could be interpreted regarding the more linear distribution of the PLLs of the recombinants from the ACW population (Figure 2.3). This quantitative resistance may also explain the resistant phenotype of 0804.174. Since qualitative resistance masks usually quantitative resistance (McDonald and Linde, 2002) we used the data obtained from progenies of 'ACW11303' also for single gene mapping. But also, if all individuals would have been excluded being offspring of 'ACW11303' the genetic window of the FB resistance locus would be in the analyzed region ranging from Ch03e03/t16k15 to rp16k15 which is covered by the sequence of BAC 16k15. Furthermore, the two individuals defining the genetic window where the resistance was mapped (0806_79 and 04208_211) were both clearly resistant with 0% PLL.

Using the data from all 33 recombinants and the deduced phenotypic data from the non-recombinant individuals enabled single gene mapping of the position of the FB-resistance locus between rp16k15 and FEM14/FEM47. This position confirmed the position of the previously published QTL and dislocated the position of the single gene analysis in the same study, which was located 9 cM distally of Ch03e03 (Peil et al., 2007). The nearest flanking markers of the resistance locus FB_MR5 (FEM14/FEM47 and rp16k15) were 0.23 cM distant from each other, which is comparable to other chromosome landing approaches for apple. Galli et al. (2010) described a distance of 0.5 cM between flanking markers (GmTNL1 and ARGH17) of the Rvi15 scab resistance locus and supplemental two co-segregating markers (41A24T7 and 43M10RP) were developed. The flanking markers developed in the study of the QTL in 'Evereste', ChFbE01 and ChFbE08, were 0.18 cM apart and marker ChFbE02-07 was co-segregating with the resistance locus (Parravicini et al., 2011). We did not identify any co-segregating marker in the present study useful for marker-assisted breeding; however, the use of two flanking markers located very close to the locus of interest is highly precise. If both flanking markers are evaluated to be in coupling with the trait locus, the trait locus will be present completely with the probability of 5.29×10^{-6} of a false positive due to double crossing over in 0.23 cM.

BAC sequencing

Hybridization with Ch03e03 identified a BAC (16k15) spanning the resistance region. The assembly of sequenced 454-pyrosequencing data of

16k15 plus an additional gap closure step via Sanger sequencing produced four contigs. Additional re-sequencing of the same BAC using Illumina technology (HiSeq 2000) could not close the remaining gaps either (data not shown), but provided additional evidence confirming the previously generated sequences. The resulting number of nucleotides (162 kbp) in the four contigs was almost equal compared to the number of base pairs of BAC 16k15, measured via PFGE and fingerprinting, which indicated that the gaps were relatively small. The region spanned by BAC 16k15 was also compared to the corresponding section of GD WGS (Velasco et al., 2010). This section ranged from marker t16k15 to EH034548 and contained approximately 23% gaps (meta-contigs, without real sequenced data, Figure 2.6), which indicated that the region was hardly sequenceable. Due to these results and the experience that non-sequenceable regions were mostly non-coding homo-polymeric stretches or repetitive elements such as SSRs we ended the investigations to complete the BAC sequence.

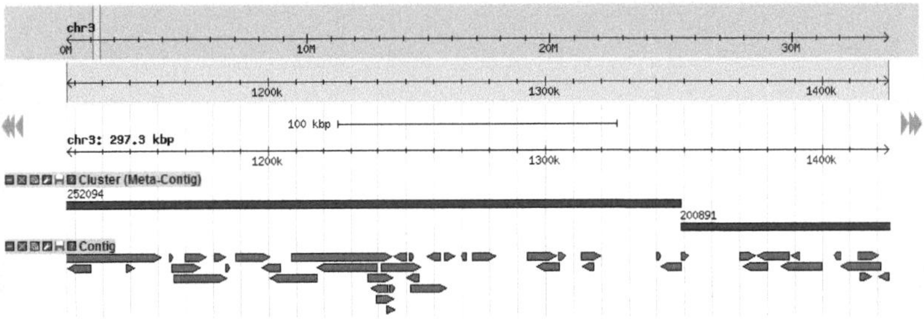

Figure 2.6 Screenshot of the *Malus* × *domestica* genome browser showing a section of LG 3 (equal to chromosome 3, chr3) of 'Golden Delicious' corresponding to the region in MR5 covered by BAC 16k15. Approximately 23% of this region was not covered by real sequenced data. The part of LG 3 presented is terminated by contigs containing t16k15 (the left side) and EH034548 (the right side) with an estimated size of 297 kb. The FB_MR5 flanking marker of MR5 rp16k15 was not present in GD WGS. EH034548 was not located at 16k15 but, as the closest marker to the centromere from rp16k15, it was chosen to flank the interval in GD. Blue: the calculated meta-contig; red: the contigs with real sequencing data; M: mega bases; k: kilo bases. Colour picture at http://e-collection.library.ethz.ch/view/eth:6104

Gene predictions

Few conclusions can be made on the gene density recorded in this work. Genes are not thought to spread equally within the genome (Barakat et al., 1999; Bernardi, 2004) so it is difficult to compare densities between different genomic regions. For example, in 48.6 kb in a GMAL 2473-derived BAC clone, Galli et al. (2010) found 17 ORFs, three of which were predicted to be related to resistance. Regarding FB resistance, Parravicini et al. (2011) predicted a total of 23 genes in 78 kb of the ornamental apple cultivar 'Evereste' using the FGENESH*tomato* algorithm; 8 of these 23 genes were related to resistance. The number of genes predicted in our study (47 genes in 162 kb) was, on the one hand, in accordance with the latter two publications, although the number of *R* gene candidates with only one candidate, *FB_MR5*, was very low. On the other hand, the GD sequence predicted 0.78 genes per 10 kb (gene density), i.e. just one quarter of what we found (Velasco et al., 2010).

However, it is noteworthy that resistance genes are usually found in clusters of *R* genes and *R* gene analogs (*RGA*s) (Meyers et al., 2003). This is also true for many *R* genes and *RGA*s of apple (Baldi et al., 2004; Calenge et al., 2005; Broggini et al., 2009). In our case, we identified a sole *R* gene that was not embedded in a cluster of paralogs. This may be explained by the lack of long-term coevolution between *Malus* species and *E. amylovora* since *E. amylovora* came from North America and *Malus* species have their origin in Central Asia (Juniper and Mabberley, 2006).

LRR-like structure

In total, 23 hydrophobic LRR-like motifs were found to fit the consensus 'LxxLx[IL]xxCxxLxxL'. Furthermore, L could be substituted by C and I due to their shared characteristics: aliphatic (I and L) as well as hydrophobic and non-polar (I, C, L) (Kyte and Doolittle, 1982; Taylor, 1986). The HHpred analysis (Söding, 2005) detected a strong similarity between the secondary structure (E-value < 1.1E-36) of FB_MR5 and the proteins 3rgz_A (brassinosteroid insensitive protein, *A. thaliana*) and 1ziw_A (TOLL-like receptor 3, *H. sapiens*). Both are folded into a horseshoe-shaped tertiary structure like classic LRRs (Kobe and Deisenhofer, 1994), suggesting that the same structure is formed by the LRR-like protein part. In addition, the protein homology modeler 'SWISS-MODEL' (Arnold et al., 2006) indicated a horseshoe-shaped protein for the LRR-like part of FB_MR5 (Figure 2.7). The number of LRR-like motifs is in accordance with the number of LRRs in other described CNLs: Galli et al. (2010) found 15, 15 and 29 imperfect LRRs in three Rvi15 (Vr2) resistance protein candidates, respectively. Parravicini et al. (2011) found 11-12 imperfect LRRs in MdE-EaN, a CNL of *Malus* cultivar 'Evereste'. The functionality of the proteins in these latter two studies has not yet been validated. Examples of functional resistance proteins of less related species possess between 14 (RPS2, *A. thaliana*) and 27 (L6, *Linum usitatissimum*) LRRs (Jones and Jones, 1997), respectively. These models and facts support the putative functionality of the LRR-like motif.

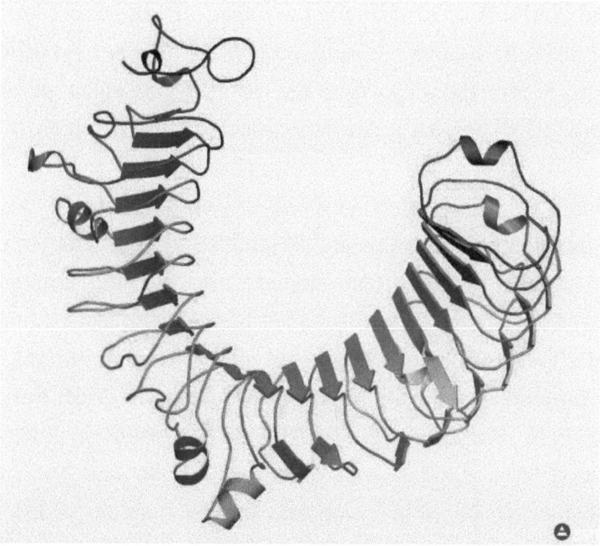

Figure 2.7 Protein model of the LRR-like structure of FB_MR5 modelled by SWISS-MODEL. The crystal structure of 3rgzA was used as the basis to model the tertiary structure. Colour picture at http://e-collection.library.ethz.ch/view/eth:6104

Assumed mode of function of FB_MR5

FB_MR5 was classified as belonging to the family of CNL proteins. The members of this protein family are often resistance genes. In the plant resistance gene database 55 CNLs are listed; 26 of these act against fungi, 10 against oomycetes, 7 against viruses, 6 against bacteria, 5 against nematodes and one against aphids (Sanseverino et al., 2010). The six CNLs that act against bacteria are Bs2, Prf, RPM1, RPS2, RPS5 and Xa1. Although Bs2, RPM1 and Xa1 were not reported as CNLs but as NBS-LRRs without a coiled-coil region (Grant et al., 1995; Yoshimura et al., 1998; Tai et al., 1999). Therefore, only Prf, RPS2 and RPS5 can be directly compared to FB_MR5. These three proteins confer the resistance following the decoy/guard model being in the position of the guard. The Prf of tomato is known as the guard of Pto, a serine-threonine kinase that recognizes

AvrPto and AvrPtoB of *P. syringae* pv. *tomato* (Pedley and Martin, 2003). The RPS2 from *A. thaliana* is activated by the effector AvrRpt2 via the cleavage of RIN4, the decoy/guardee of RPS2 (Caplan et al., 2008). Furthermore, RIN4 is also a decoy/guardee of RPM1, which is activated after phosphorylation of RIN4 by AvrRpm1 or AvrB. The RPS5 guards the protein kinase PBS1, which is AvrPphB susceptible (Caplan et al., 2008). The corresponding Avr proteins are all secreted to host cells via type three secretion systems (T3SS). This implies also for the pathosystem *E. amylovora/Malus* an interaction between host *R* gene and pathogen T3SS or T3SS-delivered effector as suggested also for the fire blight candidate resistance proteins of the cultivar 'Evereste' the kinase MdE-EaK7 and the CNL MdE-EaN (Parravicini et al., 2011). The authors hypothesized that they function in a similar system to that of Pto and Prf of tomato. Inferential, the two possible hypotheses for the function of FB_MR5 are that (1) the effectors or the T3SS itself could be targets of the candidate resistance gene product FB_MR5; or (2) that FB_MR5 has an undiscovered decoy/guardee that recognizes homologs of the aforementioned effectors of *E. amylovora*.

No conclusion about the mode of function of FB_MR5 could be made from a ClustalW pairwise alignment with Prf, RPS2 and RPS5, where 20-22% identity was calculated between the aa sequences. However, the search for effectors in *E. amylovora* encourages the second hypothesis: No Avr products corresponding to Pto and PBS1 homologs of MR5 were discovered in *E. amylovora* by BLAST search of WGS of *E. amylovora* strains ATCC 49946 and CFBP1430. Instead homolog of AvrRpt2 of *P. syringae*, AvrRpt2$_{EA}$, was identified in the WGS of *E. amylovora*. Previous studies showed that AvrRpt2$_{EA}$ elicits an HR (hypersensitive response) in *A. thaliana* after infiltration with *P. syringae* pv. *tomato* DC3000 expressing *avrRpt2*$_{EA}$ (Zhao et al., 2006). Also, *avrRpt2*$_{EA}$- strains had a significantly lower effect on pear fruits (Zhao et al., 2006). In the constitutional transcriptome of unchallenged MR5 (Fahrentrapp et al., unpublished) we also identified a homolog of RIN4, the RPS2 decoy/guardee. RIN4 homologs were also found on chromosome 5 and 10 of GD and we amplified putative homolog sequences with similar size from cultivars 'Idared' and 'Galaxy'. For these reasons we suggest that the three proteins FB_MR5, RIN4_MR5 and $AvrRpt2_{EA}$ could act in a similar 'decoy'/'guard' system to RPS2, RIN4 and AvrRpt2.

Conclusion

The FB_MR5 CNL is the second CNL found in apple related to fire blight resistance and the third candidate resistance gene of *Rosaceae* against *E. amylovora* to be expressed and cloned. The FB_MR5 CNL comprises a new LLR-like motif with the consensus LxxLx[IL]xxCxxLxxL putatively involved in protein-protein interaction, like classic LRRs. The results of the comparison to known CNLs and of the search for interacting proteins indicate that the mode of function of FB_MR5 could be in congruence with the decoy/guard model together with RIN4_MR5_and $AvrRpt2_{EA}$. Future tasks are (1) to provide evidence of the functionality of the designated *R* gene in complementation assays and (2) to describe the mode of interaction between *E. amylovora* and *M.* × *robusta* 5, including (3) clarifying the function of the LRR-like motif within FB_MR5 as well as (4) how the resistance was overcome by the two Canadian *E. amylovora* strains. Furthermore, if the functionality of the *R* gene can be confirmed, the flanking markers of *FB_MR5* could be used for highly precise MAS in classical breeding. The markers would also enable the pyramiding of different FB resistance genes and other important apple diseases such as scab and powdery mildew. The isolated *R* gene could be used to transfer the trait in well-established apple cultivars and for engineering cisgene apple plants.

Acknowledgments

We acknowledge the Genetic Diversity Center of ETH Zurich, Switzerland for sequencing, fragment analysis and bioinformatics support, as well as LeRoux P-M, Malnoy M and Baumgartner I for technical support. For financial funding we thank the Federal Office for Agriculture FOAG of Switzerland (project: ZUEFOS) as well as the D-A-CH (German-Austrian-Swiss project: 310030L_130811).

References

Altschul, S. F., Gish, W., Miller, W., Myers, E. W. and Lipman, D. J. (1990). "Basic local alignment search tool". J. Mol. Biol. 215(3): 403-410.

Arnold, K., Bordoli, L., Kopp, J. and Schwede, T. (2006). "The SWISS-MODEL workspace: A web-based environment for protein structure homology modelling". Bioinformatics 22(2): 195-201.

Bai, J., Pennill, L. A., Ning, J., Lee, S. W., Ramalingam, J., Webb, C. A., Zhao, B., Sun, Q., Nelson, J. C., Leach, J. E. and Hulbert, S. H. (2002). "Diversity in nucleotide binding site-leucine-rich repeat genes in cereals". Genome Res. 12(12): 1871-1884.

Bailey, T. L. and Elkan, C. (1994). "Fitting a mixture model by expectation maximization to discover motifs in biopolymers". Proc. Int. Conf. Intell. Syst. Mol. Biol. 2: 28-36.

Baldi, P., Patocchi, A., Zini, E., Toller, C., Velasco, R. and Komjanc, M. (2004). "Cloning and linkage mapping of resistance gene homologues in apple". Theor. Appl. Genet. 109(1): 231-239.

Barakat, A., Han, D. T., Benslimane, A.-A., Rode, A. and Bernardi, G. (1999). "The gene distribution in the genomes of pea, tomato and date palm". FEBS Lett. 463(1-2): 139-142.

Bernardi, G. (2004). "The organization of plant genomes". In "New comprehensive biochemistry". G. Bernardi. Amsterdam, Elsevier. 37: 227-240.

Block, A. and Alfano, J. R. (2011). "Plant targets for *Pseudomonas syringae* type III effectors: Virulence targets or guarded decoys?". Curr. Opin. Microbiol. 14(1): 39-46.

Broggini, G. A. L., Galli, P., Parravicini, G., Gianfranceschi, L., Gessler, C. and Patocchi, A. (2009). "*HcrVf* paralogs are present on linkage groups 1 and 6 of *Malus*". Genome 52(2): 129-138.

Broggini, G. A. L., Le Cam, B., Parisi, L., Wu, C., Zhang, H. B., Gessler, C. and Patocchi, A. (2007). "Construction of a contig of BAC clones spanning the region of the apple scab avirulence gene *AvrVg*". Fungal Genet. Biol. 44(1): 44-51.

Calenge, F., Van der Linden, C. G., Van de Weg, E., Schouten, H. J., Van Arkel, G., Denance, C. and Durel, C. E. (2005). "Resistance gene analogues identified through the NBS-profiling method map close to major genes and QTL for disease resistance in apple". Theor Appl Genet 110(4): 660-668.

Caplan, J., Padmanabhan, M. and Dinesh-Kumar, S. P. (2008). "Plant NB-LRR immune receptors: From recognition to transcriptional reprogramming". Cell Host Microbe 3(3): 126-135.

Chagne, D., Carlisle, C., Blond, C., Volz, R., Whitworth, C., Oraguzie, N., Crowhurst, R., Allan, A., Espley, R., Hellens, R. and Gardiner, S. (2007). "Mapping a candidate gene (MdMYB10) for red flesh and foliage colour in apple". BMC Genomics 8(1): 212.

DeYoung, B. J. and Innes, R. W. (2006). "Plant NBS-LRR proteins in pathogen sensing and host defense". Nat. Immunol. 7(12): 1243-1249.

Durel, C. E., Denance, C. and Brisset, M. N. (2009). "Two distinct major QTL for resistance to fire blight co-localize on linkage group 12 in apple genotypes 'Evereste' and *Malus floribunda* clone 821". Genome 52(2): 139-147.

FAO (2009). "Food and agricultural organization of the United Nations Statistical database." Retrieved Aug. 9th, 2011, from http://faostat.fao.org.

Flor, H. (1971). "Current status of gene-for-gene concept". Annu. Rev. Phytopathol. 9: 275.

Frey, J. E., Frey, B., Sauer, C. and Kellerhals, M. (2004). "Efficient low-cost DNA extraction and multiplex fluorescent PCR method for marker-assisted selection in breeding". Plant Breed. 123(6): 554-557.

Galli, P., Patocchi, A., Broggini, G. A. L. and Gessler, C. (2010). "The *Rvi15* (*Vr2*) apple scab resistance locus contains three TIR-NBS-LRR genes". Mol. Plant. Microbe Interact. 23(5): 608-617.

Gardiner, S., Norelli, J., de Silva, N., Fazio, G., Peil, A., Malnoy, M., Horner, M., Bowatte, D., Carlisle, C., Wiedow, C., Wan, Y., Bassett, C., Baldo, A., Celton, J.-M., Richter, K., Aldwinckle, H. and Bus, V.

(2012). "Putative resistance gene markers associated with quantitative trait loci for fire blight resistance in Malus 'Robusta 5' accessions". BMC Genet. 13(1): 25.

Gessler, C. (2011). "Cisgenic disease resistant apples: A product with benefits for the environment, producer and consumer". Outlooks on Pest Management 22(5): 216-219.

Grant, M., Godiard, L., Straube, E., Ashfield, T., Lewald, J., Sattler, A., Innes, R. and Dangl, J. (1995). "Structure of the *Arabidopsis* RPM1 gene enabling dual specificity disease resistance". Science 269(5225): 843-846.

Hammond-Kosack, K. E. and Jones, J. D. (1997). "Plant disease resistance genes". Annu. Rev. Plant Physiol. Plant Mol. Biol. 48: 575-607.

Jones, D. A. and Jones, J. D. G. (1997). "The role of leucine-rich repeat proteins in plant defences". In "Adv. Bot. Res.". I. C. T. J.H. Andrews and J. A. Callow, Academic Press. 24: 89-167.

Juniper, B. E. and Mabberley, D. J. (2006). "The story of the apple". Portland, Or., Timber Press.

Kajava, A. V. and Kobe, B. (2002). "Assessment of the ability to model proteins with leucine-rich repeats in light of the latest structural information". Protein Sci. 11(5): 1082-1090.

Kellerhals, M., Székely, T., Sauer, C., Frey, J. and Patocchi, A. (2009). "Pyramiding scab resistances in apple breeding ". Erwerbs-Obstbau 51(1): 21-28.

Khan, M., Zhao, Y. F. and Korban, S. (2011). "Molecular mechanisms of pathogenesis and resistance to the bacterial pathogen *Erwinia amylovora*, causal agent of fire blight disease in Rosaceae". Plant Mol. Biol. Report. 10.1007/s11105-011-0334-1: 1-14.

Kobe, B. and Deisenhofer, J. (1994). "The leucine-rich repeat: A versatile binding motif". Trends Biochem. Sci. 19(10): 415-421.

Kruijt, M., De Kock, M. J. D. and De Wit, P. J. G. M. (2005). "Receptor-like proteins involved in plant disease resistance". Mol. Plant Pathol. 6(1): 85-97.

Kyte, J. and Doolittle, R. F. (1982). "A simple method for displaying the hydropathic character of a protein". J. Mol. Biol. 157(1): 105-132.

Liebhard, R., Gianfranceschi, L., Koller, B., Ryder, C. D., Tarchini, R., Van De Weg, E. and Gessler, C. (2002). "Development and characterisation of 140 new microsatellites in apple (*Malus* x *domestica* Borkh.)". Mol. Breed. 10: 217-241.

Lupas, A. (1996). "Prediction and analysis of coiled-coil structures". In "Methods Enzymol.". F. D. Russell. Waltham, Academic Press. 266: 513-525.

Lupas, A., Van Dyke, M. and Stock, J. (1991). "Predicting coiled coils from protein sequences". Science 252(5009): 1162-1164.

Martin, G., Brommonschenkel, S., Chunwongse, J., Frary, A., Ganal, M., Spivey, R., Wu, T., Earle, E. and Tanksley, S. (1993). "Map-based cloning of a protein kinase gene conferring disease resistance in tomato". Science 262(5138): 1432-1436.

McDonald, B. A. and Linde, C. (2002). "Pathogen population genetics, evolutionary potential, and durable resistance". Annu. Rev. Phytopathol. 40(1): 349-379.

McManus, P. S., Stockwell, V. O., Sundin, G. W. and Jones, A. L. (2002). "Antibiotic use in plant agriculture". Annu. Rev. Phytopathol. 40(1): 443-465.

Meyers, B. C., Kozik, A., Griego, A., Kuang, H. and Michelmore, R. W. (2003). "Genome-wide analysis of NBS-LRR–encoding genes in *Arabidopsis*". Plant Cell 15(4): 809-834.

Ngugi, H. K., Lehman, B. and Madden, L. V. (2011). "Multiple treatment meta-analysis of products evaluated for control of fire blight in the eastern United States". Phytopathology 101(5): 512-522.

Norelli, J. L. and Aldwinckle, H. S. (1986). "Differential susceptibility of *Malus* spp cultivars Robusta-5, Novole, and Ottawa-523 to *Erwinia-Amylovora*". Plant Dis. 70(11): 1017-1019.

Norelli, J. L., Aldwinckle, H. S. and Beer, S. V. (1988). "Virulence of *Erwinia amylovora* strains to *Malus* sp. Novole plants grown in vitro and in the greenhouse". Phytopathology 78(10): 1292-1297.

Norelli, J. L., Farrell, R. E., Bassett, C. L., Baldo, A. M., Lalli, D. A., Aldwinckle, H. S. and Wisniewski, M. E. (2009). "Rapid transcriptional response of apple to fire blight disease revealed by cDNA suppression subtractive hybridization analysis". Tree Genetics & Genomes 5(1): 27-40.

Pagni, M., Ioannidis, V., Cerutti, L., Zahn-Zabal, M., Jongeneel, C. V., Hau, J., Martin, O., Kuznetsov, D. and Falquet, L. (2007). "MyHits: improvements to an interactive resource for analyzing protein sequences". Nucleic Acids Res. 35(Web Server issue): W433-437.

Parravicini, G. (2010). "Candidate genes for fire blight resistance in apple cultivar 'Evereste'". IBZ Plant Pathology, ETH, Zurich, Doctor of Sciences, Document number 19203, pages: 149

Parravicini, G., Gessler, C., Denance, C., Lasserre-Zuber, P., Vergne, E., Brisset, M. N., Patocchi, A., Durel, C. E. and Broggini, G. A. L. (2011). "Identification of serine/threonine kinase and nucleotide-binding site-leucine-rich repeat (NBS-LRR) genes in the fire blight resistance quantitative trait locus of apple cultivar 'Evereste'". Mol. Plant Pathol. 12(5): 493-505.

Patocchi, A., Vinatzer, B. A., Gianfranceschi, L., Tartarini, S., Zhang, H. B., Sansavini, S. and Gessler, C. (1999). "Construction of a 550 kb BAC contig spanning the genomic region containing the apple scab resistance gene *Vf*". Mol. Gen. Genet. 262(4): 884-891.

Paulin, J. P., Lachaud, G. and Lespinasse, Y. (1993). "Role of the aggresssivness of strains of *Erwinia amylovora* in the experimental evalutation of susceptiblility of apple cultivars to fire blight". Acta Hort. (ISHS) 338: 375-376.

Pedley, K. F. and Martin, G. B. (2003). "Molecular basis of *Pto*-mediated resistance to bacterial speck disease in tomato". Annu. Rev. Phytopathol. 41: 215-243.

Peil, A., Flachowsky, H., Hanke, M.-V., Richter, K. and Rode, J. (2011). "Inoculation of *Malus* × *robusta* 5 progeny with a strain breaking resistance to fire blight reveals a minor QTL on LG5". Acta Hort. (ISHS) 896: 357-362.

Peil, A., Garcia-Libreros, T., Richter, K., Trognitz, F. C., Trognitz, B., Hanke, M. V. and Flachowsky, H. (2007). "Strong evidence for a fire blight resistance gene of *Malus robusta* located on linkage group 3". Plant Breed. 126(5): 470-475.

Peil, A., Hanke, M. V., Flachowsky, H., Richter, K., Garcia-Libreros, T., Celton, J. M., Gardiner, S., Horner, M. and Bus, V. (2008). "Confirmation of the fire blight QTL of *Malus* x *robusta* 5 on linkage group 3". Acta Hort (ISHS) 793: 297-303.

Rairdan, G. J., Collier, S. M., Sacco, M. A., Baldwin, T. T., Boettrich, T. and Moffett, P. (2008). "The coiled-coil and nucleotide binding domains of the potato Rx disease resistance protein function in pathogen recognition and signaling". Plant Cell 20(3): 739-751.

Richter, K. and Fischer, C. (2000). "Stability of fire blight resistance in apple". Acta Hort (ISHS)(538): 267-270.

Rozen, S. and Skaletsky, H. (1999). "Primer3 on the WWW for general users and for biologist programmers". Methods Mol. Biol. 132: 365-386.

Salamov, A. A. and Solovyev, V. V. (2000). "*Ab initio* gene finding in *Drosophila* genomic DNA". Genome Res. 10(4): 516-522.

Sanseverino, W., Roma, G., De Simone, M., Faino, L., Melito, S., Stupka, E., Frusciante, L. and Ercolano, M. R. (2010). "PRGdb: A bioinformatics platform for plant resistance gene analysis". Nucleic Acids Res. 38(Database issue): D814-821.

Schouten, H. J., Krens, F. A. and Jacobsen, E. (2006). "Cisgenic plants are similar to traditionally bred plants". EMBO Rep. 7(8): 750-753.

Schuelke, M. (2000). "An economic method for the fluorescent labeling of PCR fragments". Nat. Biotechnol. 18(2): 233-234.

Söding, J. (2005). "Protein homology detection by HMM–HMM comparison". Bioinformatics 21(7): 951-960.

Szankowski, I., Waidmann, S., Degenhardt, J., Patocchi, A., Paris, R., Silfverberg-Dilworth, E., Broggini, G. A. L. and Gessler, C. (2009). "Highly scab-resistant transgenic apple lines achieved by introgression of *HcrVf2* controlled by different native promoter lengths". Tree Genetics & Genomes 5(2): 349-358.

Tai, T. H., Dahlbeck, D., Clark, E. T., Gajiwala, P., Pasion, R., Whalen, M. C., Stall, R. E. and Staskawicz, B. J. (1999). "Expression of the *Bs2* pepper gene confers resistance to bacterial spot disease in tomato". Proc. Natl. Acad. Sci. U. S. A. 96(24): 14153-14158.

Taylor, W. R. (1986). "The classification of amino acid conservation". J. Theor. Biol. 119(2): 205-218.

Van der Biezen, E. A. and Jones, J. D. (1998). "Plant disease-resistance proteins and the gene-for-gene concept". Trends Biochem. Sci. 23(12): 454-456.

Van der Hoorn, R. A. L. and Kamoun, S. (2008). "From guard to decoy: A new model for perception of plant pathogen effectors". Plant Cell 20(8): 2009-2017.

Van der Zwet, T. (2006). "Present worldwide distribution of fire blight and closely related diseases". Acta Hort. (ISHS) 704: 35-36.

Van Ooijen, G., Mayr, G., Kasiem, M. M. A., Albrecht, M., Cornelissen, B. J. C. and Takken, F. L. W. (2008). "Structure–function analysis of the NB-ARC domain of plant disease resistance proteins". J. Exp. Bot. 59(6): 1383-1397.

Van Ooijen, J. W. and Voorrips, R. E. (2001). "JoinMap 3.0, software for the calculation of genetic linkage maps". Plant Research International, Wageningen, the Netherlands.

Vanblaere, T., Szankowski, I., Schaart, J., Schouten, H., Flachowsky, H., Broggini, G. A. L. and Gessler, C. (2011). "The development of a cisgenic apple plant". J. Biotechnol. 154(4): 304-311.

Velasco, R., Zharkikh, A., Affourtit, J., Dhingra, A., Cestaro, A., Kalyanaraman, A., Fontana, P., Bhatnagar, S. K., Troggio, M., Pruss, D., Salvi, S., Pindo, M., Baldi, P., Castelletti, S., Cavaiuolo, M., Coppola, G., Costa, F., Cova, V., Dal Ri, A., Goremykin, V., Komjanc, M., Longhi, S., Magnago, P., Malacarne, G., Malnoy, M., Micheletti, D., Moretto, M., Perazzolli, M., Si-Ammour, A., Vezzulli, S., Zini, E., Eldredge, G., Fitzgerald, L. M., Gutin, N., Lanchbury, J., Macalma, T., Mitchell, J. T., Reid, J., Wardell, B., Kodira, C., Chen, Z., Desany, B., Niazi, F., Palmer, M., Koepke, T., Jiwan, D., Schaeffer, S., Krishnan, V., Wu, C., Chu, V. T., King, S. T., Vick, J., Tao, Q., Mraz, A., Stormo, A., Stormo, K., Bogden, R., Ederle, D., Stella, A., Vecchietti, A., Kater, M. M., Masiero, S., Lasserre, P., Lespinasse, Y., Allan, A. C., Bus, V., Chagne, D., Crowhurst, R. N., Gleave, A. P., Lavezzo, E., Fawcett, J. A., Proost, S., Rouze, P., Sterck, L., Toppo, S., Lazzari, B., Hellens, R. P., Durel, C.-E., Gutin, A., Bumgarner, R. E., Gardiner, S. E., Skolnick, M., Egholm, M., Van de Peer, Y., Salamini, F. and Viola, R. (2010). "The genome of the domesticated apple (*Malus* x *domestica* Borkh.)". Nat. Genet. 42(10): 833-839.

Walker, J. E., Saraste, M., Runswick, M. J. and Gay, N. J. (1982). "Distantly related sequences in the alpha- and beta-subunits of ATP synthase, myosin, kinases and other ATP-requiring enzymes and a common nucleotide binding fold". EMBO J. 1: 945-951.

Yoshimura, S., Yamanouchi, U., Katayose, Y., Toki, S., Wang, Z.-X., Kono, I., Kurata, N., Yano, M., Iwata, N. and Sasaki, T. (1998). "Expression of *Xa1*, a bacterial blight-resistance gene in rice, is induced by bacterial inoculation". Proc. Natl. Acad. Sci. U. S. A. 95(4): 1663-1668.

Zhao, Y. F., He, S. Y. and Sundin, G. W. (2006). "The *Erwinia amylovora avrRpt2(EA)* gene contributes to virulence on pear and *AvrRpt2(EA)* is recognized by *Arabidopsis RPS2* when expressed in *Pseudomonas syringae*". Mol. Plant. Microbe Interact. 19(6): 644-654.

Chapter 3

The Transcriptome of *Malus* × *robusta* 5: General Annotation and Mapping to the Fire Blight Resistance Locus on LG 3

Abstract

Fire blight is one of the most important diseases in apple orchards. Recently *FB_MR5*, a candidate resistance gene was identified in the wild apple *Malus* × *robusta* 5 (MR5) by positional cloning. The genomic sequence of 162 kb spanning the *FB_MR5 locus* was subjected to gene prediction in order to identify putative transcripts. The software FGENESH trained on *Arabidopsis*, *Vitis* and tomato sequences was used generating from 25 to 47 putative genes which were eventual annotated using BLAST. Besides the reported CC-NBS-LRR gene *FB_MR5*, no further putative resistance gene was identified. In each predicted set of genes at least one gene was confirmed by the transcriptome data produced with next-generation sequencing of mRNA of unchallenged MR5 leaves. Transcriptomic data were also compared to the 'Golden Delicious' gene set, matching about 50% of the genes. The 'Golden Delicious' homolog genes of MR5 with expression levels >800 RPKM were then annotated: two thirds of the homolog genes expressed in MR5 are involved in metabolic and cellular processes. In addition we tried to establish a qPCR assay for measuring the transcription of *FB_MR5* in relation to ubiquitin as a reference gene in representative progenies of MR5. The method revealed high variability in the expression of the reference gene as well as the target gene *FB_MR5*, indicating also several false negative results among technical replicates. Generally *FB_MR5* was constitutively expressed in the resistant individuals and it was not expressed in the susceptible ones. The expression measured with qPCR was similar to the expression of *FB_MR5* versus ubiquitin measured in the transcriptome analysis. Nonetheless further research is needed to optimize qPCR conditions.

Introduction

In many apple orchards worldwide the most important bacterial disease is fire blight (FB) which is caused by the Gram-negative bacterium *Erwinia amylovora*. High natural resistance is found only in wild and ornamental apple species. Only so called moderately resistant cultivars such as 'Arlet' and 'Cox's Orange Pippin' are commercially available (Koski and Jacobi, 2009). One of the strongest natural resistance sources was found by means of QTL analysis on linkage group (LG) 3 of the wild apple *Malus* × *robusta* 5 (MR5) (Peil et al., 2007). This QTL on top of LG 3 explained 80% of the phenotypic variation of resistance which ranged from 0-100 percent lesion length (PLL) in the cross 'Idared' × MR5 (Peil et al., 2007). The region of the LG carrying the genetic basis of resistance was spanned by a bacterial artificial chromosome (BAC) whose sequencing by next generation sequencing methods resulted in four separate contigs (Chapter 2). On the sequence of this BAC *FB_MR5*, a candidate resistance gene against fire blight, was identified. *FB_MR5* codes for a CNL (coiled-coil nucleotide binding site leucine rich repeat protein). It was predicted with the software FGENESH trained on tomato sequences. Different other softwares for gene prediction are available such as GlimmerHMM, GeneWise and Twinscan (Salamov and Solovyev, 2000; Korf et al., 2001; Birney et al., 2004; Majoros et al., 2004) but FGENESH was evaluated as prediction software producing the most accurate results on *Drosophila melanogaster* and maize sequences (Salamov and Solovyev, 2000; Yao et al., 2005). The FGENESH algorithm has been trained on different organisms and is available for plants trained on the organisms *Arabidopsis*, *Medicago*, corn, rice, wheat, barley, *Nicotiana tabacum*, tomato, *Vitis vinifera* and *Hevea*. Today, no prediction software is available that is trained specifically for the *Rosaceae* family. The FGENESH algorithm trained on tomato sequences was previously used to identify putative transcripts from apple derived genomic sequences (Parravicini, 2010). In addition to the FGENESH tomato algorithm the gene prediction on the nucleotide sequence derived from the MR5 BAC carrying *FB_MR5*, gene sets were also predicted with FGENESH algorithms trained on sequences of the wood producing *Vitis* and the intensively studied *Arabidopsis*. Depending on the prediction algorithm 25-47 mRNAs were predicted on the same sequence (Chapter 2).

In the present work the mRNAs predicted on the MR5 BAC sequence were annotated. RNA-Seq data of unchallenged MR5 was used to asses the quality of the prediction results on *Malus* derived sequences. Further comparison of the transcriptome with the 'Golden Delicious' (GD) gene set led identify the expressed homolog genes of MR5 and expressed genes of MR5 without homolog in GD. The latter are also potential candidate resistance genes due to their absence in the highly susceptible GD. In addition the expression level of *FB_MR5* calculated on RNA-Seq data were compared to the preliminary results of the expression measured with quantitative real-time PCR (qPCR) in a set of fire blight resistant and susceptible plants.

Materials and Methods

Plant material

Eight genotypes being progeny of MR5 were selected for expression analysis of *FB_MR5*. Four of the selected genotypes were classified as resistant (0802.167, 0803.111, 0806.79, 0807.1) and four as susceptible (0802.161, 0802.191, 0802.249, 0803.292; Chapter 2). The eight genotypes and additional the susceptible cultivars 'La Flamboyante', 'Galaxy' and 'Idared', the breeding number 'ACW11303' and the resistant MR5 were grafted on M9 rootstocks in 15 replicates ('Idared' in 18) and grown in the greenhouse until healthy and actively growing shoots emerged. If all graftings grew healthy 12 graftings of each genotype ('Galaxy', 4 graftings; 'Idared', 15 graftings) were used for artificial shoot inoculation for FB phenotyping (Chapter 2) and 3 grafted replicates were used as biological replicates in the quantitative real time PCR (qPCR) assay. Leave material from the three MR5 plants used for qPCR was also used for transcriptome analysis. The resistance phenotype was assessed by artificial inoculation of single shoots performed by cutting the two youngest leaves with scissors dipped into an *E. amylovora* strain Ea222_JKI suspension as described in Chapter 2. Percent lesion length (PLL) of shoots was recorded after 21 days post inoculation. Data transformation to classify the individuals as susceptible and resistant, respectively, was done as suggested by Durel et al. (2009).

RNA processing

For quantitative real time PCR one young leave of each of the three biological replicates (each replicate was represented by a single grafted plant) was harvested and directly frozen in liquid N_2. Plants were phenotypically healthy and unchallenged (non-infected, untreated) at harvest time. After cell disruption RNA was extracted (Concert™ Plant RNA Reagent, Invitrogen, Carlsbad, CA, U.S.) treated with DNase (TURBO DNA-free™ Kit, Applied Biosystems, Carlsbad, CA, U.S.) and 12.5 µl were transcribed into cDNA (Maxima™ Reverse Transcriptase, Fermentas, Waltham, MA, USA) as previously described (Chapter 2). The cDNA was quantified with NanoDrop 8000 (Thermo Scientific, Waltham, MA, USA). The potential contamination with genomic DNA was evaluated by applying a qPCR with UBC primers (see below) on all RNA samples which were processed with the same method used for cDNA transcription but without adding the reverse transcriptase (-RT qPCR assay).

For the transcriptome analysis 3 young leaves of MR5 plants were directly frozen in liquid N_2 and used for RNA extraction with the same kit as described above but using the 'large scale' procedure. Quality and quantity was estimated with Bioanalyzer (Agilent Technologies, Santa Clara, CA, USA). The RNA was shipped to BGI (Hong Kong) accompanied with dry ice. It was sequenced from a 200 bp insert library using Illumina sequencing technology (Illumina, HiSeq 2000, San Diego, CA, USA) and SOAPdenovo (Li et al., 2009) was used for assembly. Further handling of the reads as well as the assembled data was done with CLC Genomics Workbench 4.8 (CLC Bio, Aarhus, Denmark).

BAC sequencing, mRNA prediction and comparison

The BAC 16k15, spanning the fire blight resistance locus, was sequenced and open reading frames (ORFs) predicted as described in Chapter 2. The three predicted sets of ORFs were named and abbreviated as *Arabidopsis*-, Tomato- and *V. vinifera*-based-prediction (Adp, Tbp and Vbp). The derived sequences were annotated using BLAST search against nucleotide collection of NCBI as well as BLASTx search against non-redundant UniProtKB/SwissProt database of NCBI. Their function was further analyzed with blast2GO for their gene ontology (Conesa et al., 2005). The three on the BAC of MR5 predicted gene sets were compared to each other in terms of number of mRNAs sharing the same predicted structure and the positions of TSS (transcriptional start), start codons and stop codons as well as for the number of predicted exons. The existence of shared TSS as well as start and stop codons gives a qualitative portrait on the level of the diversity of the predicted gene sets. Unique or distinct intron-exon boundaries were identified with CLC Genomics Workbench 4.8 (CLC Bio, Aarhus, Denmark). The predicted gene sets were visualized with software 'Argo' (www.broadinstitute.org).

Transcriptome vs predicted genes of MR5

The transcriptome derived sequencing reads were mapped with the 'RNA-Seq' algorithm of CLC Genomics Workbench 4.8 against the BAC sequence which was annotated with the three predicted ORFs sets (Abp, Vbp and Tbp). Intron-exon boundaries distinguishable between the gene sets were selected and the transcriptomic reads spanning the exon-exon junctions of these intron-exon boundaries were used as portion to quantify which of the prediction algorithms produced the most reliable results. Expression levels of genes were calculated as RPKM (Mortazavi et al., 2008).

Transcriptome vs 'Golden Delicious' gene set

The transcriptome derived reads of MR5 were mapped with the RNA-Seq algorithm of CLC Genomics Workbench 4.8 against the *M. x domestica* gene set (http://genomics.research.iasma.it) which is based on the whole genome sequence of cultivar 'Golden Delicious' (Velasco et al., 2010). The expression of homolog genes was calculated as RPKM values. All sequences with an expression RPKM of more than 800 were analyzed for their putative function with blast2GO annotating the sequences with GOs (gene ontologies). Additional the identified unigenes of MR5 were mapped with the 'Large gap read mapping' plugin of CLC Genomics Workbench 4.8 against the GD gene set. The unmapped unigenes with sizes > 999 bp were selected and analyzed for additional putative resistance genes against FB perfoming BLASTx against NCBI's non-redundant UniProtKB/SwissProt sequences.

Quantitative real time PCR

Expression levels of *FB_MR5* in the different genotypes were quantified relatively to ubiquitin expression levels using quantitative real time PCR (qPCR) with '5x Hot FirePol® EvaGreen® qPCR Mix Plus' (Solis BioDyne, Tartu, Estonia) on a '7500 Fast Real-Time PCR System' (Applied Biosystems, Carlsbad, CA, U.S.). Reactions were carried out in 10 µl volumes with 80 nM primers and 1 µl of 1:5 diluted cDNA. The cycling protocol included a 15 minutes 95°C step followed by 50 cycles of

15 seconds 95°C and 60 seconds 60°C. Melting curve was applied after each run from 60-95°C with a ramp time of 30 minutes. The primer efficiency (E) of used primers FB_MR5q1 and ubiquitin (UBC) (Table 3.1) was determined by amplifying the fragment in five technical repetitions from MR5 cDNA dilution series in 1:10 dilution steps from 1 to 10^{-7}. The expression of *FB_MR5* was analyzed in three biological replicates (r1-r3) and three technical replicates. Calculation of primer efficiency and expression levels of *FB_MR5* relative to ubiquitin were performed according to Pfaffl (2001) taking into account the Ct values only up to cycle 40 due to observed unspecific amplification after cycle 40. The threshold line was set manually above background noise to 0.09 delta RN.

Table 3.1 Sequences of primers used to amplify fragments of ubiquitin and *FB_MR5*.

Primer	Forward primer	Reverse primer	Amplicon size	E[e] in %
FB_MR5q1[a]	TTTATGGAGAGTGCTCCTTGC	AGCGAATCAAGGTTCTCTGG	187[c]	117[f]
UBC[b]	CGAATTTGTCCGAAGGCGT	CAATGATTGTCACAGCAGCCA	57[d]	94

[a]this study. [b]Vanblaere (2011). [c]in MR5. [d]in apple cultivar 'Fuji'. [e]efficiency of qPCR calculated on the slope of Ct values derived from amplification of a cDNA dilution series. [f]calculated on three dilution steps and repeated with three different cDNA templates

Results

Fire blight phenotyping

The recorded percent lesion lengths (PLLs) caused by *E. amylovora* on the eight selected genotypes and their parents ranged from 0 to 25% (Table 3.2). Four individuals were classified as resistant (average PLL: 4%) and four as susceptible (average PLL: 16%) which was in congruence to their genotypic data (Chapter 2). MR5 showed no necrosis (0% PLL) after inoculation, 'Idared' had a PLL of 52% and 'Galaxy' 77%.

Table 3.2 Fire blight phenotype evaluation of the selected genotypes and their parents. Plants were scissor-inoculated with *E. amylovora* strain Ea222_JKI.

Individual number	Cross	Number of replicates	Average PLL[a] in %	Phenotypic classification[b]	Genotypic classification[b]
0802.167	'ACW11303' × DA02_2,7[c]	12	5	R	R
0803.111	'ACW11303' × DA02_2,40[c]	12	3	R	R
0806.79	'La Flamboyante' × DA02_1,27[c]	12	0	R	R
0807.1	'La Flamboyante' × DA02_2,40[c]	12	7	R	R
0802.161	'ACW11303' × DA02_2,7[c]	11	14	S	S
0802.191	'ACW11303' × DA02_2,7[c]	12	15	S	S
0802.249	'ACW11303' × DA02_2,7[c]	12	12	S	S
0803.292	'ACW11303' × DA02_2,40[c]	12	25	S	S
'ACW11303'	'ACW6104' × 'Rewena'	11	12	S	S
'La Flamboyante'	N/A	7	27	S	S
'Galaxy'	N/A	4	77	S	N/A
'Idared'	N/A	15	52	S	S
MR5	*M. baccata* × *M. prunifolia*	10	0	R	R

[a]Percent lesion length; [b]R, resistant; S, susceptible; [c]resistant progeny of the cross 'Idared' × MR5 (Chapter 2)

Gene prediction and analysis

A total of 25, 41 and 47 ORFs were predicted with the algorithm trained on *V. vinifera* (Vbp), *Arabidopsis* (Abp) and tomato (Tbp) with an average mRNA length of 1485, 1065 and 942 bp (Table 3.3). The positions of exons on the four contigs of BAC 16k15 are shown in Figure 3.1. Six mRNAs were predicted identically in the three sets of predicted genes. Overall 14 TSS positions are shared between predicted gene sets. No additional common TSS of Vbp and Tbp were found but 13 between Abp/Tbp and 7 between Abp/Vbp (Figure 3.2A). 3, 5, and 2 TSS positions were not predicted in the set Abp, Tbp and Vbp, respectively. 16 start codon positions were predicted equally in the three sets as well as 15 stop codons (Figure 3.2B and C). The start and stop codons of the remaining predicted mRNAs, respectively, had no homolog or only homologs in one of the two other prediction sets. The Abp and Tbp sets shared 9 start and 12 stop codons. Abp and Vbp shared 5 start and 5 stop codons whereas Vbp and Tbp shared no start and no stop codon.

Table 3.3 General numbers of predicted gene sets compared to the transcriptome of MR5. Genes were predicted with different FGENESH algorithms on the sequence of BAC 16k15 spanning the fire blight resistance region.

Gene set	Number of predicted genes	Total bp of mRNA	Average number of exons per mRNA	Genes with at least one paired read with 100% identity	Number of mRNAs that are at least 80% covered with MR5 transcriptome reads[a]
Abp	41	37119	3.9	12	4
Tbp	47	44256	3.2	12	5
Vbp	25	43686	5.2	12	3

Vbp, *V. vinifera* based prediction; Tbp, Tomato based prediction; Abp, *Arabidopsis* based prediction.[a]analyzed with the RNA-Seq algorithm

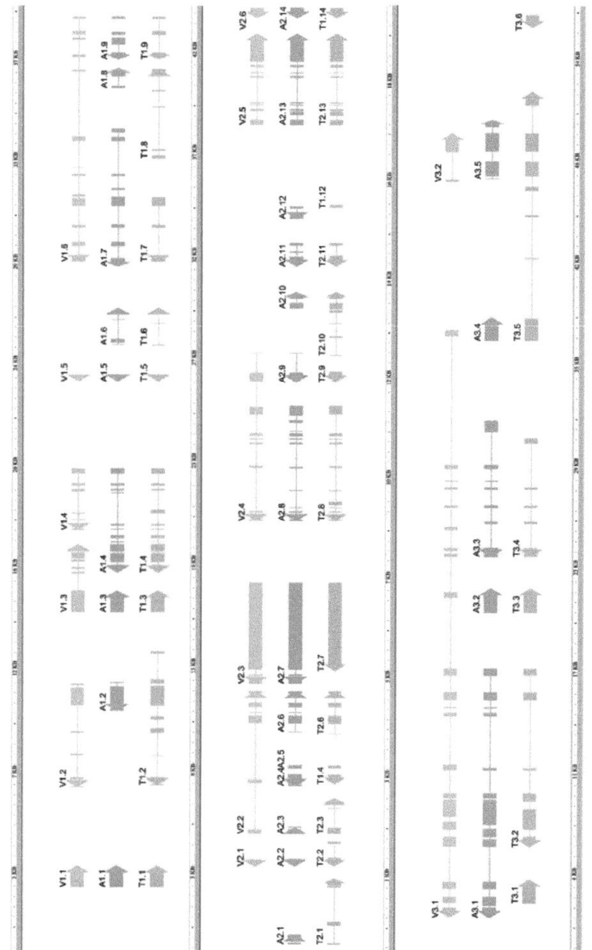

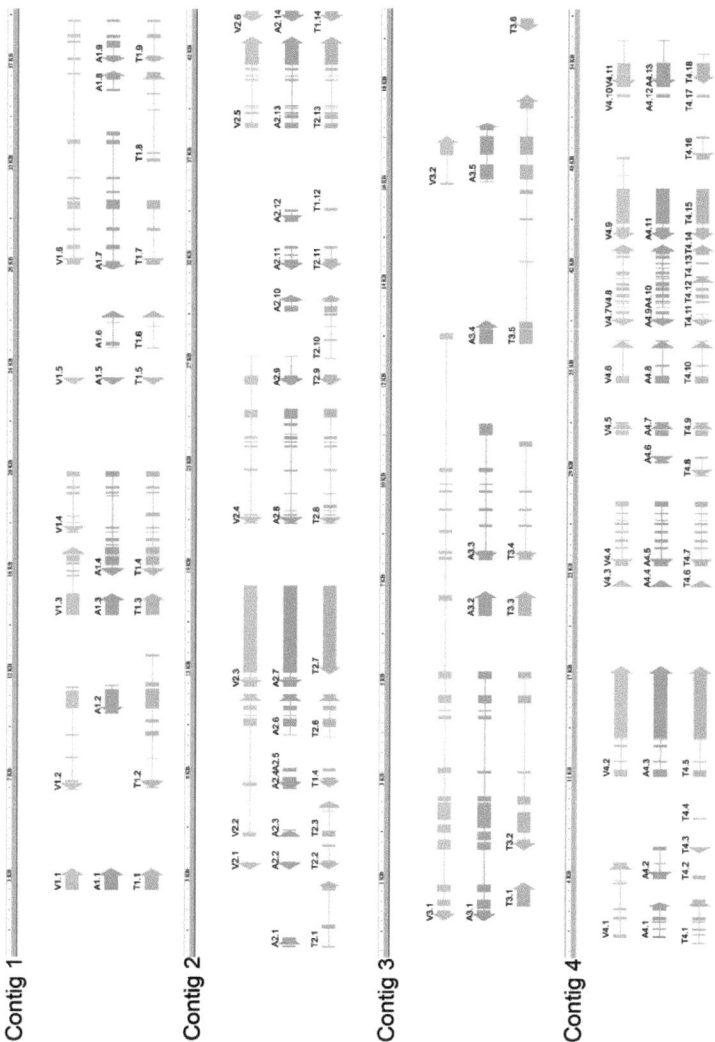

Figure 3.1 Three predicted ORF sets positioned on the four contigs of BAC 16k15. BAC 16k15 spanned the region harboring the candidate *R* gene *FB_MR5* of MR5 against *E. amylovora* (Chapter 2). The ORFs were predicted with FGENESH trained on *V. vinifera* (V, yellow, upper line), *Arabidopsis* (A, blue, middle line) and tomato (T, green, lower line). Colour picture at http://e-collection.library.ethz.ch/view/eth:6104

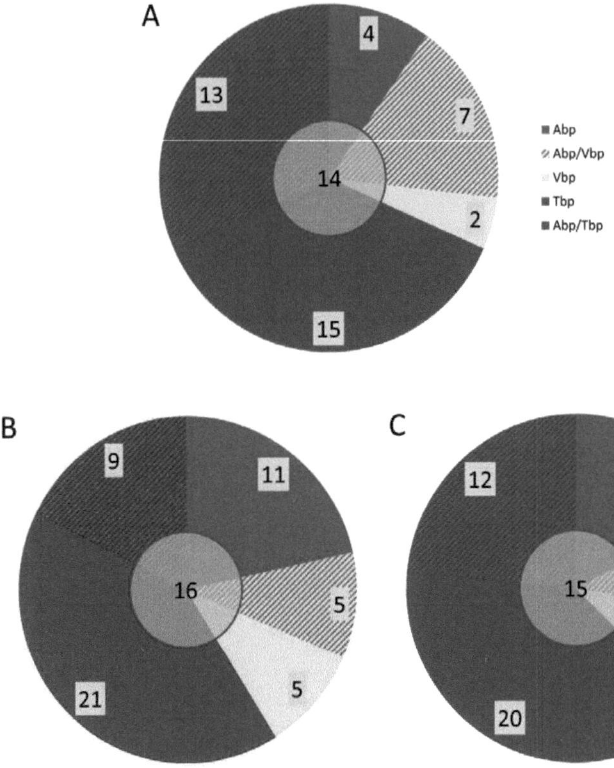

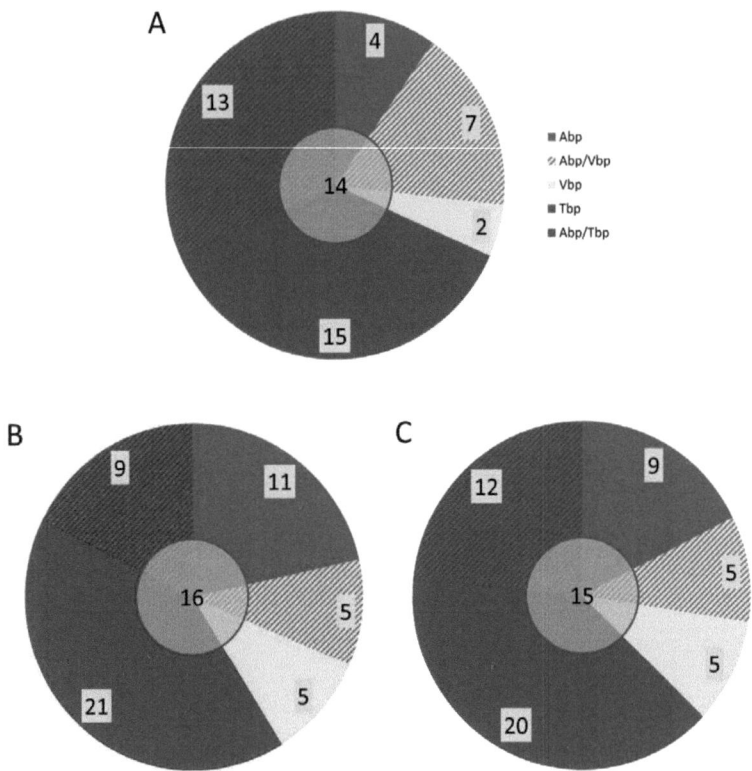

Figure 3.2 Overlapping of mRNAs predicted on the sequence of BAC 16k15 with FGENESH trained on *Arabidopsis* (Abp), *V. vinifera* (Vbp) and tomato (Tbp). Given values indicate the number of TSS positions (A), start (B) and stop (C) codons, respectively, of Abp (blue), Vbp (yellow) and Tbp (red) and the shared numbers of TSS, start and stop codons of Abp/Vbp (blue-yellow striped) and Abp/Tbp (blue-red striped). The transparent circles indicate the number of codons present in all prediction sets. No shared codons were found between Tbp and Vbp predictions except the codons shared by all prediction sets. 3, 5, and 2 TSS positions were not predicted in the set Abp, Tbp and Vbp, respectively. Colour picture at http://e-collection.library.ethz.ch/view/eth:6104

Annotation of predicted genes

Using nucleotide BLAST 44% (Vbp), 55% (Tbp) and 56% (Abp) of the predicted gene sets, respectively, were not homolog to any sequence in the nucleotide collection database. The most often found sequence homolog to the predicted mRNAs (V1.6, V4.8, T.1.4, T.1.8, T.1.9, T4.13, T4.14, A1.8, A1.9, A4.10) was the sequence with accession number AB270792. This sequence comprised a 317 kb sequence spanning the S locus (self-incompatibility locus) of *M.* × *domestica* and it contained two *SFBB* genes (for *S* locus *F-b*ox *b*rothers) (Sassa et al., 2007) which have not yet been related to any disease resistance function. Another often identified homolog sequence to predicted mRNAs (V2.1, V2.3, T2.2, T2.7, A2.2, A2.7) was the sequence of '*M. baccata* clone OLE1-19 NBS-LRR-like protein gene' (HQ399004) which was not further described in any publication. Its protein sequence was shown to share sequence homologies with the NB-ARC domain which is part of many resistance genes (Van der Biezen and Jones, 1998). The genes V2.3, T2.7 and A2.7 which were homolog to the latter sequence were investigated in depth and T2.7 was designated as candidate resistance gene against fire blight of MR5 (Chapter 2). The genes T4.11 and T4.12 were found to be homolog to the 'Vf apple scab resistance protein *HcrVf2*-like gene of *M. floribunda*' (EU794445). EU794445 has a size of 3299 bp (Broggini et al. 2009) and T4.11 and T4.12 were 471 and 747 bp in size, respectively. The homolog sequence parts of T4.11 and T4.12 to EU794445 were only 270 and 81 bp in size, respectively. Analyzing T4.11 and T4.12 in detail using MotifScan (Pagni et al., 2007) no motifs were identified which indicated any functionality of the two ORFs as resistance genes. Therefore we did not investigate the putative expression of T4.11 and T4.12. The other annotated ORFs shared homologies to sequences coding either for proteins described as 'predicted and putative' or for proteins such as the f-box family proteins, transposon and retrotransposon genes which are not known to act as resistance genes (for details of the nucleotide BLAST see Appendix B). Detailed annotations produced by BLASTx of the predicted genes are presented in

Table 2.3 and commented in Chapter 2.Gene ontology analysis identified nucleic acid binding and transferase activity as the two major molecular functions of the predicted genes (Figure 3.3). The transferase activity is focused mainly on the transfer of phosphorus-containing and glycosyl groups.

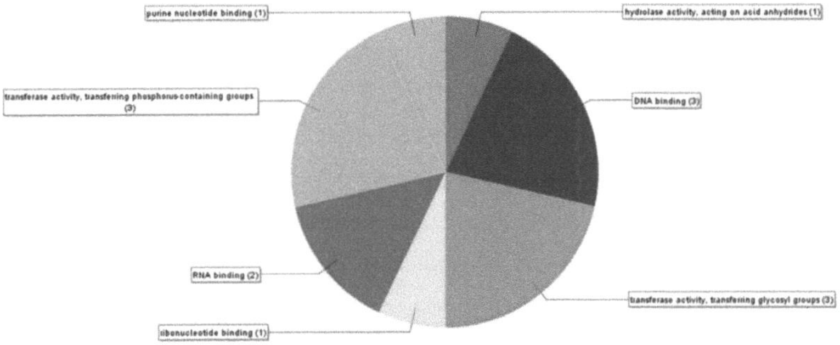

Figure 3.3 Gene ontology analysis of genes predicted on the sequence of BAC16k15. The results are classified with respect to the molecular function of the genes. The molecular function of FB_MR5 was categorized as ATP binding and the biological process defined as related to 'defense response' and 'apoptosis'. The numbers in brackets are scores given by blast 2GO being proportional to the number of sequences which are involved in the specific process. The figures were produced with blast2GO. Colour picture at http://e-collection.library.ethz.ch/view/eth:6104

Comparison: Transcriptome vs predicted genes of MR5

RNA-Seq of MR5 produced 13,400,492 reads (13,384,284 reads in pairs) with an average length of 90 nucleotides. These reads were assembled to 131,580 contigs and 61,922 unigenes were identified with lengths between 150 up to 5,178 bp (Figure 3.4).

Using the RNA-Seq algorithm 4 (Abp), 5 (Tbp) and 3 (Vbp) mRNAs were covered more than 80% with transcriptomic reads but no transcript could be fully retrieved (Table 3.3 and Appendix A). Eleven genes could be found in each gene set with at least one paired read matching completely the predicted sequence (Appendix A). Five exon-exon junctions spanned with transcriptomic reads with differently predicted splicing profiles were identified (Table 3.5). In total four exon-exon junctions predicted with FGENESH *Vitis*, two predicted with FGENESH *Arabidopsis* and one predicted with FGENESH tomato were accurately predicted. The expression of *FB_MR5* in the transcriptome is 440 RPKM and 4.4 times higher than the expression of ubiquitin.

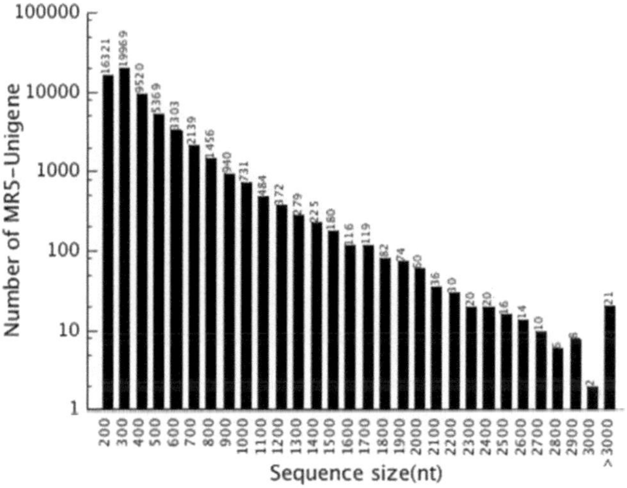

Figure 3.4 Length distribution of unigenes of MR5. In total 61,922 unigenes were identified in the transcriptome of unchallenged MR5. Lengths are given in nucleotides (nt).

Table 3.5 Proof of predicted genes by selected exon-exon junction spanning reads

Exon-exon junction	Abp	Vbp	Tbp
#1	1[a]	1[a]	0
#2	0[b]	1	0[b]
#3	0	1[c]	1[c]
#4	0	1	0
#5	1	0	0
Sum	2	4	1

[abc]pairs with same splicing profile

Comparison: Transcriptome vs Golden Delicious gene set

31,958 genes were identified in the MR5 transcriptome to be at least partially homolog to genes of the GD gene set with a total of 63,541 genes. The expression values of the homolog genes ranged from 0.05 to 36,640 RPKM. 137 genes with an RPKM of at least 800 were selected. Blast2GO functional annotation of the selected genes concerning their function in biological processes identified about two thirds of the sequences being involved in metabolic and cellular processes. Another fraction of about one eighth was related to response to stimulus. The remaining genes were related to different inter- and intracellular processes as shown in Figure 3.5A. Analyzing the GO terms concerning their molecular function, more than half of the genes are supposed to be involved in binding of e.g. proteins, nucleotides and lipids and one third is supposed to have catalytic activity (Figure 3.5B).

Mapping the unigenes of MR5 transcriptome against the GD gene set identified 17,208 unigenes without homolog. Selecting the unigenes with a length >999 bp no gene with putative function in resistance against pathogens could be identified.

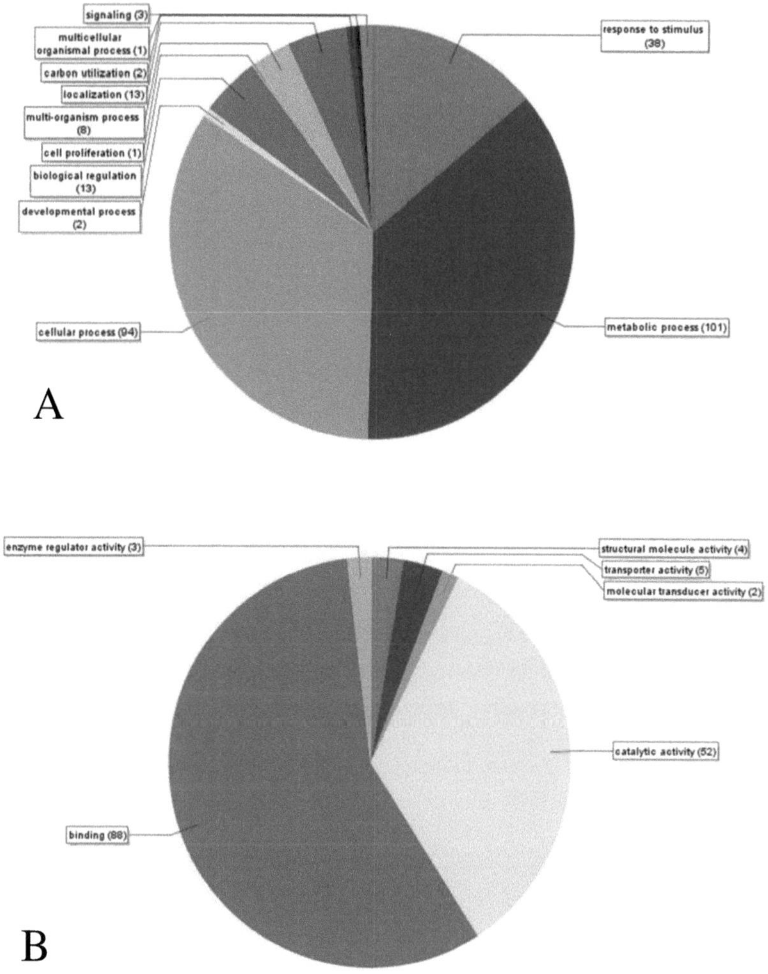

Figure 3.5 Gene ontology analysis of 137 expressed genes of MR5. These genes were found to be homolog to genes of 'Golden Delicious' and with an expression of more than 800 RPKM. A, with blast2GO sorted GO concerning to the general biological processes in which they are involved. B, the GO of the genes evaluated in terms of their molecular function. The numbers in brackets are scores given by blast 2GO being proportional to the number of sequences which are involved in the specific process. The

figures were produced with blast2GO. Colour picture at http://e-collection.library.ethz.ch/view/eth:6104

Quantitative real time PCR assay and in planta expression of FB_MR5

The preliminary results of the qPCR assay showed a detection limit for UBC and FB_MR5q1 amplification of 465 fg and 3 ng, respectively. Therefore the primer efficiency of FB_MR5q1 was calculated instead of a complete dilution series (diluted from 1 to 10^{-7}) with slopes calculated from three independent dilution series from 1 to 10^{-2}. The three dilution series were made using three different cDNA templates of three independent RNA extractions of MR5. The resulting efficiencies of FB_MR5q1 were averaged for expression level calculations. The efficiency of the two primer pairs was calculated with the slope of Ct values plotted against the logarithm of the concentration resulting in 117% (R^2=0.99) and 94% (R^2=0.99) for FB_MR5q1 and UBC, respectively. The melting temperature of amplified fragments estimated after each qPCR run was 82.5 and 77.5°C of FB_MR5q1 and UBC, respectively. Therefore, putative primer dimers with a melting temperature of about 75°C for FB_MR5q1 should be distinguishable from the target product. The difference of melting temperature of UBC amplified fragments *versus* primer dimers represented only 1.5°C, hence, they were only distinguishable if the amplification plots and the derivatives of the melt curves of positive and no template controls as well as the samples were compared on screen. In UBC qPCR reactions primer dimers occurred always in reactions without target template. In FB_MR5q1 reactions no primer dimers were observed. Unspecific amplification occurs after cycle 40 in reactions with primers FB_MR5q1. Therefore we only included cycle 1-40 in the analysis.

Subject to verification we could state that *FB_MR5* was expressed in MR5 and in the four tested resistant progenies. In the resistant progeny 0802.167 only one of the three biological replicates expressed *FB_MR5*. The expression levels of *FB_MR5* relative to ubiquitin varied between 0.02

(0807.1-r1) and 21 (0807.1-r3) with an average of 7.4 taking into account the individuals which expressed *FB_MR5* (Figure 3.6). MR5 had an average expression of 6. In all resistant individuals at least one biological replicate showed amplification of FB_MR5q1, however, a total of five out of 15 failed, namely 0802.167-r2 and -r3, 0806.76-r1, 0803.111-r1 and MR5-r3. The susceptible parents as well as the susceptible progenies did not express the target gene. Ubiquitin was expressed in all biological and technical replicates of resistant as well as of susceptible individuals accept of 0803.111-r1 (Figure 3.6).

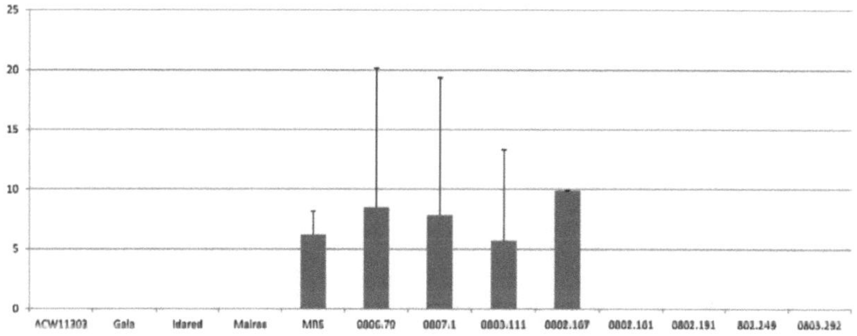

Figure 3.6 Expression of *FB_MR5* in averaged biological replicates. Expression of *FB_MR5* was only detected in MR5 itself and its resistant progenies. The susceptible parents and the susceptible progenies did not express the targeted gene. Error bars display the standard deviation of two (0806.79, 0803.111, MR5) and three replicates (0807.1), respectively. Colour picture at http://e-collection.library.ethz.ch/view/eth:6104

Discussion

Different de novo gene prediction softwares are available, but none of these softwares is trained especially for the prediction of genes of *Malus* species. The development of specific algorithms is often part of genome sequencing projects. Unfortunately no such training has been performed in relation to the first genome sequencing of *M. x domestica* cultivar 'Golden Delicious' (Velasco et al. 2010). Taking into account the positive evaluation of FGENESH we decided to predict genes on a 162 kb sequence of MR5 using FGENESH trained on different organisms. The predictions with the differently trained algorithms led to highly variable numbers of ORFs with different positions of their TSS, start and stop codons as well as their exons (Figure 3.1). Comparing the positions of TSS, start and stop codons the biggest fraction of genes predicted uniquely was always found in the Tbp. Even if the percentage of genes of each set were compared, Tbp reached 30-44% of unique genes for the three different codon positions. Vbp and Abp had unique reads between 8-20% and 10-26%, respectively. Most similarities concerning these three positions were found between Tbp and Abp. Interestingly Vbp and Tbp shared no additional TSS and codon positions than the ones shared by all prediction sets. But as detailed below FGENESH *Vitis* and tomato are the two most promising algorithms in terms of predicting true gene variants. In BLASTx of the predicted genes 46 different hits were identified (Chapter 2). 83% of them were only predicted or hypothetical proteins. Therefore the functional annotation was supported by gene ontology analysis which identified transferase activity and nucleic acid binding as the two major molecular functions of the predicted proteins (Fig 3.3).

Eleven predicted genes from each set were represented by at least one matching transcriptome derived read. These genes are therefore supposed to be transcribed at least partially and with a different splicing profile, respectively. Only three BLASTx hits could be associated with these predicted genes: COBL7 (COBRA-LIKE 7), GPI mannosyltransferase 2 and F-box/kelch-repeat protein At1g55270. The family of F-box/kelch-repeat proteins is supposed to control the "degradation of cellular regulatory proteins via the ubiquitin pathway" (Bai et al., 1996; Andrade et al., 2001). The COBRA protein family is involved in cell expansion (Roudier et al., 2005). The GPI mannosyltransferase 2 transfers the second mannosyl of the glycosylphosphatidylinositol pathway, that anchors proteins to the cell surface (Kang et al., 2005). According to the function of the homolog hits the eleven predicted genes are not supposed to be involved in resistance towards bacterial pathogens.

Comparing the transcriptome of MR5 generated by high throughput sequencing of RNA isolated from unchallenged leaves with the three predicted gene sets should indicate which of the algorithms is most suited. It is very likely that low transcribed genes are not covered by any reads, indicating a) a low depth of sequencing b) the low density of trancriptome reads to validate splicing profiles. The latter was possible only for 5 exon-exon junction predicted differently in at least two gene sets. In this way the Vbp was confirmed four times, the Abp two times and the Tbp only once. Based on our results the *Vitis* predicted gene set showed the closest number of predicted genes compared to the genes confirmed by RNA sequencing, and also the best potential for detecting correct exon-exon junctions. But the true number of genes and their splicing profile is not finally determinable with the presented data set. Summarized, taking also into account that *FB_MR5* was predicted in the true version by FGENESH tomato (Chapter 2) we suggest to use FGENESH trained on tomato and *Vitis* to predict genes in *Malus*. Deeper RNA sequencing, also following infection should allow increasing the reliability of the evinced results and identifying also up or down regulated genes that may play a role during defense.

Expression of FB_MR5

The expression of *FB_MR5* was evaluated *in planta* and in transgene calli (Chapter 4) using qPCR. All qPCR results produced in this study should be handled with reservation due to high variability between biological replicates. *FB_MR5* was expressed in MR5 and in the resistant progeny of MR5. In two biological replicates of 0802.167 (r2 and r3) no amplification was detected. The Ct values of UBC amplification in these individuals was measured with 33 and 35, respectively, proving the presence of cDNA; FB_MR5q1 was amplifiable from genomic DNA of these replicates; and genotypic validation of the three replicates with the molecular marker Ch03e03 showed the allele coupled with the resistance of MR5 (data not shown). This suggested that either the method needs to be optimized or *FB_MR5* was not or only incompletely transcribed. Therefore we suggest optimizing the qPCR assay as detailed below.

A high variability of expression levels in between biological replicates was detected. This made it impossible to compare the phenotypic resistance levels to qPCR results of *FB_MR5* expression. The high variability of expression levels may be caused by the variation of the concentration of the template cDNA, unstable expression or amplification of target and reference, respectively. As shown in Figure 3.6 the input cDNA concentration ranged from 250ng/ul to 770ng/ul using the same methodological processing for all samples. This variability is commonly normalized with the use of a reference gene. As reference gene a constantly expressed gene such as elongation factor, 18S, actin and others should be used (Nicot et al., 2005). In qPCR experiments with apple elongation factor, RUBISCO, 28S, beta-tubulin, ubiquitin and actin were used as reference genes (Paris et al., 2009; Szankowski et al., 2009; Cova et al., 2010; Joshi et al., 2011; Vanblaere, 2011). We used ubiquitin as reference gene because in pre-experiments no amplification could be observed of the other potential reference gene primers tested. In our experimental setup the expression of the reference gene over time and plant age was not taken into account because the plants rose under same conditions and RNA was extracted from healthy and unchallenged plant material. The Ct values of the reference UBC as well as the target FB_MR5q1 showed no correlation to the input template concentration (R^2 of 0.02 and 0.05, Figure 3.7) and therefore indicated that further optimization of the qPCR conditions is essential. Therefore the main result of this experiment is of qualitative nature. *FB_MR5* was expressed in resistant progenies of MR5 and MR5 itself and it was not expressed in any of the susceptible plants.

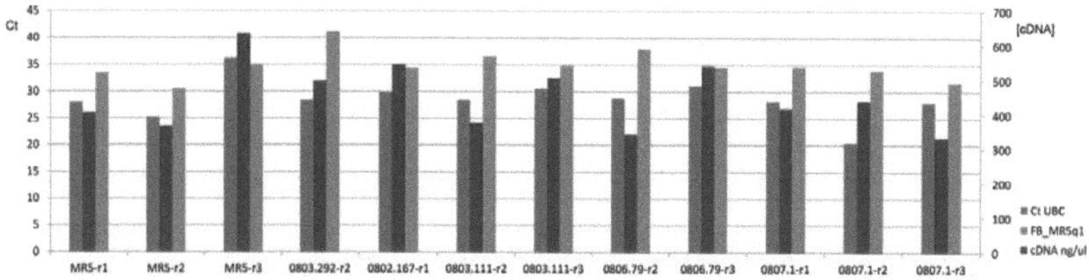

Figure 3.7 cDNA template concentration compared to UBC and FB_MR5q1 threshold cycles (Ct) of quantitative real time PCR. The template concentration of cDNA of each sample is given in ng/ul (dark red columns) measured with NanoDrop. The Ct value of amplification of UBC and FB_MR5q1, amplifiying fragments of ubiquitin and FB_MR5, respectively, are given in blue and orange columns. Colour picture at http://e-collection.library.ethz.ch/view/eth:6104

The preliminary expression values of *FB_MR5* of averaged MR5 replicates showed a relatively small standard deviation (6.2±1.9). Comparing the latter expression from qPCR to the expression measured in the transcriptome analysis (4.4 times higher as ubiquitin), we detected comparable values that need to be verified in further experiments.

Further optimization of the qPCR assay is indispensable. The optimization process should fulfill the MIQE (Bustin et al., 2009) guidelines including a strong evaluation of a set of reference genes, different RNA extractions and cDNA transcription kits as well as qPCR systems/kits (e.g. EvaGreen®, SybrGreen®, TaqMan®). Possible reference gene candidates could also be identified by RNA-Seq of different tissues grown under different conditions such as temperature, UV exposure and pathogen infection.

Conclusion

The prediction of genes on the genomic sequence of MR5 derived from top of linkage group 3 was done using FGENESH. Since there is no *Malus* specifically trained FGENESH available we compared the algorithms trained on *Arabidopsis*, *Vitis* and tomato with the transcriptome of MR5. The results indicate a better accuracy for the algorithm trained on tomato and *Vitis* than *Arabidopsis*. *FB_MR5* was found to be expressed in MR5 itself and in tested resistant progenies of MR5 and it was not expressed in the susceptible progenies. The surrounding genes of *FB_MR5* were annotated with nucleotide BLAST and BLASTx without identifying any new candidate fire blight resistance genes. The high expressed MR5 genes homolog to GD genes belonged mainly to genes of cellular and metabolic processes and are mostly involved in binding and catalytic activity. Future works should include the in-depth validation of gene prediction algorithms specifically for *Malus* species and the validation of a set of reference genes for quantitative real time PCR.

References

Andrade, M. A., González-Guzmán, M., Serrano, R. and Rodríguez, P. L. (2001). "A combination of the F-box motif and kelch repeats defines a large Arabidopsis family of F-box proteins". Plant Mol. Biol. 46(5): 603-614.

Bai, C., Sen, P., Hofmann, K., Ma, L., Goebl, M., Harper, J. W. and Elledge, S. J. (1996). "SKP1 Connects Cell Cycle Regulators to the Ubiquitin Proteolysis Machinery through a Novel Motif, the F-Box". Cell 86(2): 263-274.

Birney, E., Clamp, M. and Durbin, R. (2004). "GeneWise and Genomewise". Genome Res. 14(5): 988-995.

Bustin, S. A., Benes, V., Garson, J. A., Hellemans, J., Huggett, J., Kubista, M., Mueller, R., Nolan, T., Pfaffl, M. W., Shipley, G. L., Vandesompele, J. and Wittwer, C. T. (2009). "The MIQE Guidelines: Minimum Information for Publication of Quantitative Real-Time PCR Experiments". Clin. Chem. 55(4): 611-622.

Conesa, A., Götz, S., García-Gómez, J. M., Terol, J., Talón, M. and Robles, M. (2005). "Blast2GO: a universal tool for annotation, visualization and analysis in functional genomics research". Bioinformatics 21(18): 3674-3676.

Cova, V., Paris, R., Passerotti, S., Zini, E., Gessler, C., Pertot, I., Loi, N., Musetti, R. and Komjanc, M. (2010). "Mapping and functional analysis of four apple receptor-like protein kinases related to *LRPKm1* in *HcrVf2* -transgenic and wild-type apple plants". Tree Genetics & Genomes 6(3): 389-403.

Durel, C. E., Denance, C. and Brisset, M. N. (2009). "Two distinct major QTL for resistance to fire blight co-localize on linkage group 12 in apple genotypes 'Evereste' and *Malus floribunda* clone 821". Genome 52(2): 139-147.

Joshi, S., Schaart, J., Groenwold, R., Jacobsen, E., Schouten, H. and Krens, F. (2011). "Functional analysis and expression profiling of *HcrVf1* and *HcrVf2* for development of scab resistant cisgenic and intragenic apples". Plant Mol. Biol. 75(6): 579-591.

Kang, J. Y., Hong, Y., Ashida, H., Shishioh, N., Murakami, Y., Morita, Y. S., Maeda, Y. and Kinoshita, T. (2005). "PIG-V Involved in Transferring the Second Mannose in Glycosylphosphatidylinositol". J. Biol. Chem. 280(10): 9489-9497.

Korf, I., Flicek, P., Duan, D. and Brent, M. R. (2001). "Integrating genomic homology into gene structure prediction". Bioinformatics 17(suppl 1): S140-S148.

Koski, R. D. and Jacobi, W. R. (2009). "Fire Blight." Retrieved January, 27th, 2012, from http://www.ext.colostate.edu/index.html.

Li, R., Zhu, H., Ruan, J., Qian, W., Fang, X., Shi, Z., Li, Y., Li, S., Shan, G., Kristiansen, K., Li, S., Yang, H., Wang, J. and Wang, J. (2009). "De novo assembly of human genomes with massively parallel short read sequencing". Genome Res. 10.1101/gr.097261.109.

Majoros, W. H., Pertea, M. and Salzberg, S. L. (2004). "TigrScan and GlimmerHMM: two open source ab initio eukaryotic gene-finders". Bioinformatics 20(16): 2878-2879.

Mortazavi, A., Williams, B. A., McCue, K., Schaeffer, L. and Wold, B. (2008). "Mapping and quantifying mammalian transcriptomes by RNA-Seq". Nat. Methods 5(7): 621-628.

Nicot, N., Hausman, J.-F., Hoffmann, L. and Evers, D. (2005). "Housekeeping gene selection for real-time RT-PCR normalization in potato during biotic and abiotic stress". J. Exp. Bot. 56(421): 2907-2914.

Pagni, M., Ioannidis, V., Cerutti, L., Zahn-Zabal, M., Jongeneel, C. V., Hau, J., Martin, O., Kuznetsov, D. and Falquet, L. (2007). "MyHits: improvements to an interactive resource for analyzing protein sequences". Nucleic Acids Res. 35(Web Server issue): W433-437.

Paris, R., Cova, V., Pagliarani, G., Tartarini, S., Komjanc, M. and Sansavini, S. (2009). "Expression profiling in *HcrVf2* -transformed apple plants in response to *Venturia inaequalis*". Tree Genetics & Genomes 5(1): 81-91.

Parravicini, G. (2010). "Candidate genes for fire blight resistance in apple cultivar 'Evereste'". IBZ Plant Pathology, ETH, Zurich, Doctor of Sciences, Document number 19203, pages: 149

Peil, A., Garcia-Libreros, T., Richter, K., Trognitz, F. C., Trognitz, B., Hanke, M. V. and Flachowsky, H. (2007). "Strong evidence for a fire blight resistance gene of *Malus robusta* located on linkage group 3". Plant Breed. 126(5): 470-475.

Pfaffl, M. W. (2001). "A new mathematical model for relative quantification in real-time RT-PCR". Nucleic Acids Res. 29(9): e45.

Roudier, F., Fernandez, A. G., Fujita, M., Himmelspach, R., Borner, G. H. H., Schindelman, G., Song, S., Baskin, T. I., Dupree, P., Wasteneys, G. O. and Benfey, P. N. (2005). "COBRA, an Arabidopsis Extracellular

Glycosyl-Phosphatidyl Inositol-Anchored Protein, Specifically Controls Highly Anisotropic Expansion through Its Involvement in Cellulose Microfibril Orientation". The Plant Cell Online 17(6): 1749-1763.

Salamov, A. A. and Solovyev, V. V. (2000). "*Ab initio* gene finding in *Drosophila* genomic DNA". Genome Res. 10(4): 516-522.

Sassa, H., Kakui, H., Miyamoto, M., Suzuki, Y., Hanada, T., Ushijima, K., Kusaba, M., Hirano, H. and Koba, T. (2007). "S locus F-box brothers: Multiple and pollen-specific F-box genes with S haplotype-specific polymorphisms in apple and japanese pear". Genetics 175(4): 1869-1881.

Szankowski, I., Waidmann, S., Degenhardt, J., Patocchi, A., Paris, R., Silfverberg-Dilworth, E., Broggini, G. A. L. and Gessler, C. (2009). "Highly scab-resistant transgenic apple lines achieved by introgression of *HcrVf2* controlled by different native promoter lengths". Tree Genetics & Genomes 5(2): 349-358.

Van der Biezen, E. A. and Jones, J. D. G. (1998). "The NB-ARC domain: a novel signalling motif shared by plant resistance gene products and regulators of cell death in animals". Curr. Biol. 8(7): R226-R228.

Vanblaere, T. (2011). "The development of a cisgenic scab resistant apple cv. 'Gala'". IBZ Plant Pathology, ETH Zurich, Doctor of Science, Document number pages: 122

Velasco, R., Zharkikh, A., Affourtit, J., Dhingra, A., Cestaro, A., Kalyanaraman, A., Fontana, P., Bhatnagar, S. K., Troggio, M., Pruss, D., Salvi, S., Pindo, M., Baldi, P., Castelletti, S., Cavaiuolo, M., Coppola, G., Costa, F., Cova, V., Dal Ri, A., Goremykin, V., Komjanc, M., Longhi, S., Magnago, P., Malacarne, G., Malnoy, M., Micheletti, D., Moretto, M., Perazzolli, M., Si-Ammour, A., Vezzulli, S., Zini, E., Eldredge, G., Fitzgerald, L. M., Gutin, N., Lanchbury, J., Macalma, T., Mitchell, J. T., Reid, J., Wardell, B., Kodira, C., Chen, Z., Desany, B., Niazi, F., Palmer, M., Koepke, T., Jiwan, D., Schaeffer, S., Krishnan, V., Wu, C., Chu, V. T., King, S. T., Vick, J., Tao, Q., Mraz, A., Stormo, A., Stormo, K., Bogden, R., Ederle, D., Stella, A., Vecchietti, A., Kater, M. M., Masiero, S., Lasserre, P., Lespinasse, Y., Allan, A. C., Bus, V., Chagne, D., Crowhurst, R. N., Gleave, A. P., Lavezzo, E., Fawcett, J. A., Proost, S., Rouze, P., Sterck, L., Toppo, S., Lazzari, B., Hellens, R. P., Durel, C.-E., Gutin, A., Bumgarner, R. E., Gardiner, S. E., Skolnick, M., Egholm, M., Van de Peer, Y., Salamini, F. and Viola, R. (2010). "The genome of the domesticated apple (*Malus* x *domestica* Borkh.)". Nat. Genet. 42(10): 833-839.

Chapter 3

Yao, H., Guo, L., Fu, Y., Borsuk, L. A., Wen, T.-J., Skibbe, D. S., Cui, X., Scheffler, B. E., Cao, J., Emrich, S. J., Ashlock, D. A. and Schnable, P. S. (2005). "Evaluation of five *ab initio* gene prediction programs for the discovery of maize genes". Plant Mol. Biol. 57(3): 445-460.

Chapter 4

Is *FB_MR5* a Functional Resistance Gene?

Abstract

Recently a candidate resistance gene against fire blight (FB) of *Malus* × *robusta* 5 (MR5), *FB_MR5*, coding for a CNL was identified. *FB_MR5* was cloned into an *Agrobacterium*-transferable vector under the control of CaMV 35S promoter and its own promoter, respectively. Both constructs were used to transform FB susceptible *in vitro* 'Gala' plants leading to transgenic calli of which selected samples expressed the integrated target gene. 21 putatively transgenic shoots developed and were transferred to proper elongation medium. They will be later micro propagated and micro grafted to M9 rootstocks and then tested for their resistance towards fire blight infection caused by the bacterium *Erwinia amylovora*.

Introduction

Fire blight is one of the most dangerous diseases in apple (*Malus* × *domestica*) production worldwide. The affecting bacteria, *Erwinia amylovora*, caused losses of several million US$ for example in Michigan in the year 2000 (Norelli et al. 2003). The most widely cropped apple cultivars such as 'Gala', 'Golden Delicious' and 'Fuji' are highly susceptible and chemicals, copper compounds, bio control agents as well as antibiotics are used to control the disease. Another control measure could be the application of naturally occurring resistance. FB resistance is found amongst ornamental cultivars and wild apple species and can be introgressed by conventional breeding. This is a very time-consuming, expensive and laborious task, due to high level of heterozygosity, cultivar self-incompatibility and a long generation cycle with a minimum of four years. A shorter generation duration can be achieved through the use for example with the early flowering technology (Flachowsky et al. 2011; Le Roux et al. 2011) or the so called 'low input fast-track' method (Baumgartner et al. 2011). However, breeding will always result in a new apple cultivar with its one characteristic. Hence, genetic modification and especially cisgenesis could become the strategy of choice to maintain the cultivar's characteristics and respond rapidly to producers' demand of a fire blight resistant cultivar. Cisgenesis was defined as the genetic engineering of genes derived from crossable donor plants under the control of their original promoter and terminator without any foreign genes such as overexpression promoters or antibiotic resistance genes used as selection markers (Schouten et al. 2006). The product does not contain any foreign genes that could not be introduced by conventional breeding. Therefore it might be more accepted by consumers (Gessler 2011). The application of this technique in apple was described previously using the apple scab resistance gene *HcrVf2* (Belfanti et al. 2004; Szankowski et al. 2009) which was implemented in a cisgenic 'Gala' (Vanblaere et al. 2011). In the case of FB two very strong quantitative trait loci (QTL) for resistance against

FB were detected on linkage group (LG) 12 of the ornamental apple cultivar 'Evereste' and on LG 3 of the wild apple *Malus* × *robusta* 5 explaining 68 and 80% of phenotypic variation, respectively (Durel et al. 2009; Peil et al. 2007). The phenotypic variation ranged in the cross 'Idared' × MR5 from 0-100% of necrotized shoot (Peil et al. 2007); for 'Evereste' it was not described. From 'Evereste' two candidate resistance genes *MdE-EaK7* and *MdE-EaN* have been recently isolated (Parravicini et al. 2011) but their functionality was not yet proven. The resistance locus of MR5 was studied in detail (Chapter 2) and one single candidate gene *FB_MR5* comprising the sequence for a CNL (coiled-coil nucleotide binding site leucine rich repeat) protein was isolated and cloned (Chapter 2). It was hypothesized that FB_MR5 would function in analogy to the decoy/guard model (Van der Hoorn and Kamoun 2008) together with the effector $AvrRpt2_{EA}$ (Zhao et al. 2006) and the decoy/guardee RIN4_MR5, that was detected in the transcriptome of MR5 (Chapter 5).
To provide evidence for the function of *FB_MR5* as resistance gene against FB we report a complementation assay of susceptible apple cultivar 'Gala' with *FB_MR5* under the control of its own promoter/terminator and the promoter CaMV 35S/ocs terminator, respectively. The expression of the target gene in a selection of transgene calli was analyzed with quantitative real time PCR (qPCR). The regenerated transgene plants will be inoculated with *E. amylovora* strain Ea222_JKI (Chapter 2) in following works to prove the functionality of the putative resistance gene.

Materials and Methods

Plant material and transformation

For the transformation we used 'Gala' plants derived from INRA Angers (France) and Julius-Kühn Institute (JKI Dresden-Pillnitz, Germany). The *in vitro* apple plants of cultivar 'Gala' provided by INRA Angers grew on previously described media (Vanblaere et al. 2011) and the *in vitro* 'Gala' plants provided by JKI grew on media described by Milcevicova et al. (2010) under the following regime: 16/8 (hours light/darkness), 4 kLux and 24°C. *Agrobacterium*-mediated transformation was done according to Szankowski et al. (2003) and Vanblaere et al. (2011) in four separate transformation experiments (Table 4.1). In the first and second experiment (T36 and T37) the leaves of JKI derived *in vitro* 'Gala' plants were used; in the third and fourth experiment (T40 and T41) the *in vitro* 'Gala' plants originally from INRA Angers were transformed. Two different plasmids constructed by DNA-Cloning (Hamburg, Germany) were used for transformation. The first plasmid, p95N::FB_MR5, contained the backbone p95N and inserted *FB_MR5* including its own promoter (2 kbp upstream) and terminator region (1.5 kbp downstream) (Figure 4.1 A) and was transformed in experiments T36 and T41. In the second plasmid, p9N35S::FB_MR5, *FB_MR5* was under the control of the constitutive promoter CaMV 35S and the ocs terminator (Figure 4.1 B). This plasmid was transferred in experiments T37 and T40. Both plasmids contained the sequence of the eukaryotic selection marker *npt2* coding for kanamycin resistance. After transformation calli were regenerated from explants on kanamycin containing regeneration medium (Szankowski et al. 2009) under the same light and temperature conditions as the in vitro 'Gala' plants. Calli were transferred every 14 days to fresh media until shoots developed. Regenerated shoots were transferred to selective elongation medium containing kanamycin (Szankowski et al. 2009). The medium was renewed every six weeks. DNA of putatively transgenic shoots was extracted (Frey et al. 2004) and analyzed with primers NPTII (Vanblaere 2011) and the specific primer for FB_MR5 (FB_MR5q1, Chapter 3) to evaluate if the target genes NPTII and FB_MR5 were transferred to the plants.

In future works, regenerated transgene shoots growing on elongation media will be micro grafted to M9 rootstocks and acclimatized to the greenhouse as described by Joshi (2010). Later on, the transgenic plants will be phenotyped for resistance against fire blight.

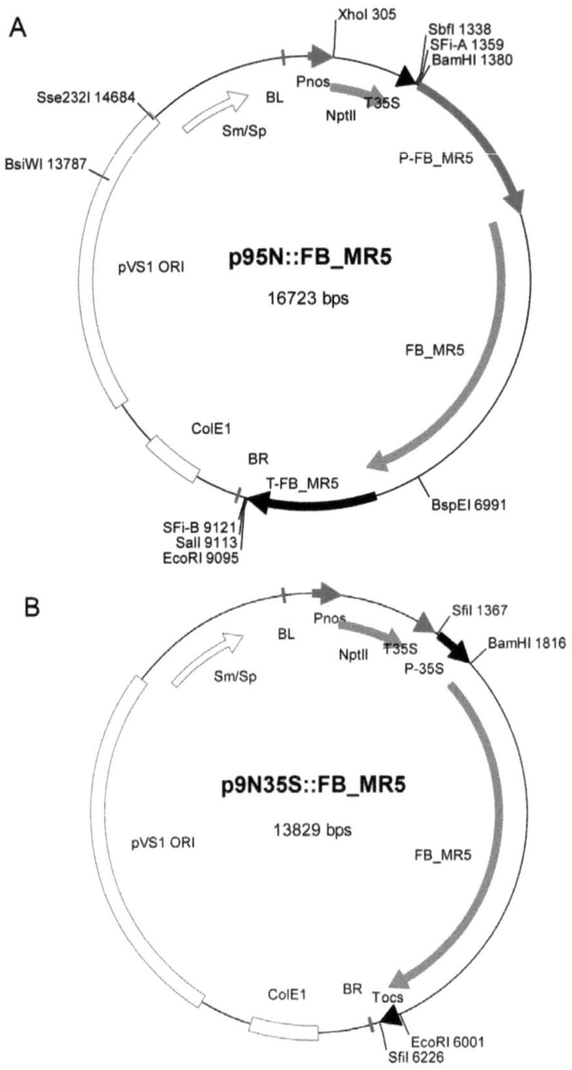

Figure 4.1 Plasmids used for *Agrobacterium*-mediated transformation of *in vitro* 'Gala' plants. A, p95N::FB_MR5 contained *FB_MR5* under the control of its own promoter (P-FB_MR5) and terminator (T-FB_MR5). B, p9N35S::FB_MR5 contained

FB_MR5 under the control of CaMV 35S promoter (P-35S) and ocs terminator (Tocs). Colour picture at http://e-collection.library.ethz.ch/view/eth:6104

RNA extraction, cDNA transcription and quantitative real time PCR

RNA was extracted according to the manufacturer's instructions (Concert™ Plant RNA Reagent, Invitrogen, Carlsbad, CA, USA) from thin tissue slides cut from two green and well growing calli of each of the transformations T36, T37 and T40 and from three calli of T41. RNA extraction was followed by DNase digestion using TURBO DNA-free™ Kit (Ambion®, Carlsbad, CA, USA).

The RNA was transcribed into to cDNA using 'Maxima™ Reverse Transcriptase' (Fermentas, Waltham, MA, USA) with oligo dT primers according to the manufacturer.

qPCR was carried out as described in Chapter 2 using specific primers for the target gene *FB_MR5* (FB_MR5q1) as well as the reference genes ubiquitin (UBC). The expression level of *FB_MR5* was determined relative to UBC and the contamination with genomic DNA was excluded in a -RT qPCR control assay (Chapter 3).

Results

Tranformation

In the transformation experiments T36 and T37 of *in vitro* 'Gala' plants 200 explants per plasmid were plated and transferred every two weeks to new selective regeneration media. Three months after transformation 30 and 60 green and putative transgene calli containing p95N::FB_MR5 and p9N35S::FB_MR5, respectively, were collected and remained actively growing on the media (Table 4.1). In the experiments T40 and T41 with plasmids p9N35S::FB_MR5 and p95N::FB_MR5, respectively, 40 and 80 green calli were obtained from originally 360 explants (Table 4.1). Eleven (T36 and T37) and seven (T40 and T41) months after the transformation experiments, respectively, 24 growing shoots regenerated from the calli which were transferred to elongation medium (Figure 4.2). 21 of them were evaluated as true transgenic shoots by PCR amplification with both primers, FB_MR5q1 and NPTII.

Table 4.1 Transformation experiment and collected number of calli.

Transformation number	Plant material	Plasmid	Number of explants	Number of calli	Number of shoots
T36	'Gala' JKI	p95N::FB_MR5	200	30	4
T37	'Gala' JKI	p9N35S::FB_MR5	200	60	0
T40	'Gala' INRA	p9N35S::FB_MR5	180	40	5
T41	'Gala' INRA	p95N::FB_MR5	180	80	15

Figure 4.2 Appearing transgenic shoots of 'Gala' harboring p9N35S::FB_MR5. The shoots are growing on elongation medium and will be propagated and grafted on M9 rootstocks for further resistance level evaluation against *E. amylovora*. Size standard corresponds to approximately 1 cm. Colour picture at http://e-collection.library.ethz.ch/view/eth:6104

Expression of FB_MR5

FB_MR5 was found to be expressed in seven out of nine samples (Figure 4.3 A). The two calli which did not express *FB_MR5* were T37 sample two (T37.2) and T40 sample one (T40.1). All nine tested calli expressed the reference gene UBC as shown in quantification plots of qPCR (Figure 4.3 B). FB_MR5q1 fragments amplified from calli melted at the same temperature (82.5 °C) as the corresponding amplicons of MR5 (Figure 4.3 C). The Ct values of FB_MR5q1 and UBC ranged from 30-35 and 20-28, respectively (Figure 4.4). The expression levels of *FB_MR5* relative to the reference gene were between 0.01 (41.3) and 1.4 (41.1; Figure 4.4).

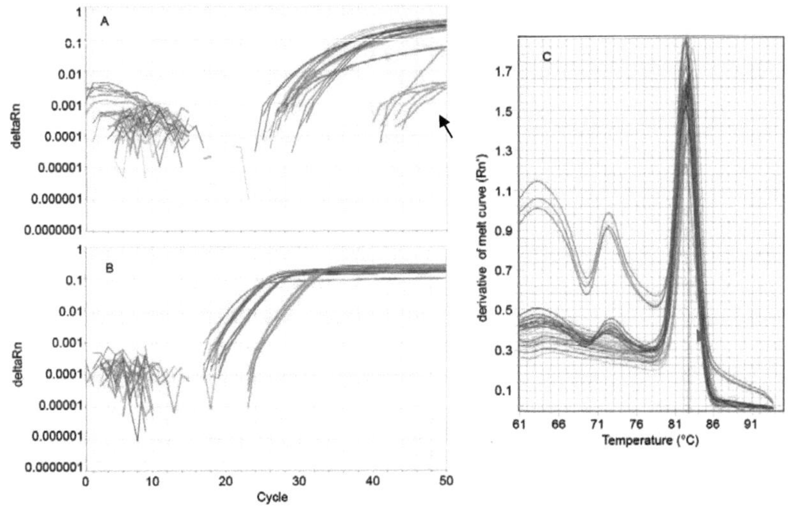

Figure 4.3 Amplification with quantitative real time PCR of FB_MR5q1 and UBC of nine callus derived cDNA samples. Increase of fluorescence of amplicons (deltaRn) amplified with FB_MR5q1 (A) and UBC (B), respectively, which were fluorescently labeled with EvaGreen®. Black arrow points to unspecific amplification of T37.2 and T40.1. C, first derivative of the melt curve of FB_MR5q1 amplified products with a maximum at 82.5°C (vertical blue line), red arrow points to melt curve of FB_MR5q1 amplified from MR5 displayed in light blue. Colour picture at http://e-collection.library.ethz.ch/view/eth:6104

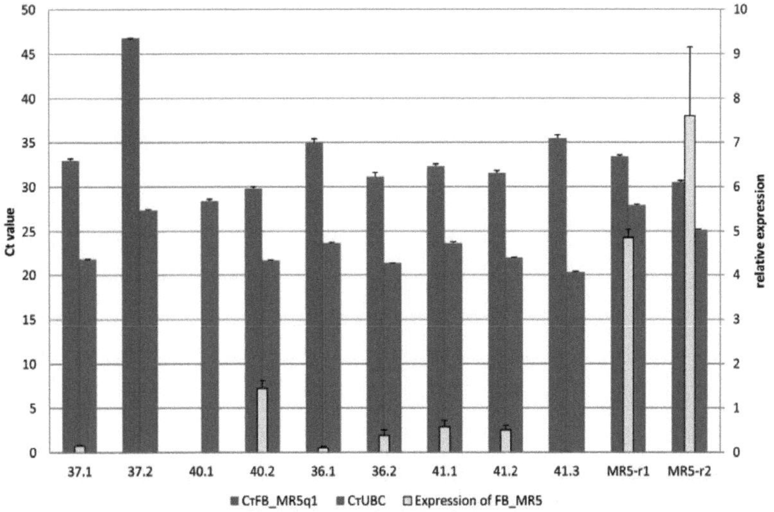

Figure 4.4 Expression of *FB_MR5* in callus tissues relative to the expression of ubiquitin and the corresponding Ct values of the amplification. Two (T36, T37 and T40) and three (T41) calli, respectively, per transformation experiment were analyzed with qPCR for their expression levels of *FB_MR5* relative to ubiquitin (yellow columns). Additional the Ct values are given of the two primer pairs FB_MR5q1 (blue columns) and UBC (red columns). MR5-r1 and MR5-r2 are two separately extracted cDNA samples of MR5 leaves. Colour picture at http://e-collection.library.ethz.ch/view/eth:6104

Discussion

Gala plants were transformed with two different constructs, one harboring the gene of interest under the control of its own promoter and one under the control of the CaMV 35S promoter. Especially in apples complementation assays are often laborious and time consuming due to the long transformation time and shoot regeneration as well as low transformation efficiency. This was reported for example by Vanblaere et al. (2011) who regenerated 10 transgenic lines out of 1635 explants and two of their transformation experiments resulted in no single regenerated shoot. The shoot regeneration appeared to be the most critical step in the transformation process because we regenerated a total of 190 actively growing green calli out of 760 explants and only 24 shoots were regenerated from the calli. The difficulty of shoot regeneration is also reflected in the number of different protocols for apple shoot regeneration from apple derived callus tissue describing the use of e.g. different auxin and cytokine derivatives in different concentrations, different sugars as well as gelrite instead of agar based growing media (Malnoy et al. 2008). The authors of more than half of the protocols we compared, suggested to incubate calli for shoot induction in darkness (Liu et al. 1983; Sriskandarajah et al. 1994; Bondt et al. 1996; Norelli et al. 1999) the others suggested to induce shoot development under light (Gamage and Nakanishi 2000; Szankowski et al. 2003). We followed a protocol based on Szankowski et al. (2003) and modified as described by Szankowski et al. (2009) and Vanblaere et al. (2011) using a 16/8 h night/day simulation. In our experiments first shoots appeared after 3-6 months after transformation and new shoots continued to develop after 11 months. In a side experiment we kept some calli in darkness but also from those shoot development was not faster.

Expression of *FB_MR5* was detected in seven out of nine selected calli in very low levels compared to the expression level in MR5 *in planta*. The expression level of the target gene in MR5 (Chapter 2) measured relative to the ubiquitin expression, was about five times higher than the expression of *FB_MR5* in the transgene calli. This could be caused by different expression levels in different tissues such as leaves (MR5) and undifferentiated calli tissue. During the development of plant and callus formation different gene expression profiles were described in *A. thaliana*. In callus tissue 241 genes were up-regulated and 373 genes were down-regulated whereas e.g. during shoot formation 478 and 397 genes were up- and down-regulated, respectively (Che et al. 2006). Another reason for the differences in expression levels between mRNA extracted from transformed calli and untransformed leaves could be that the tested calli did not express the transgene but were contaminated with the *A. tumefaciens* used for transformation which may lead to a transient expression of the target gene (Birch 1997). Transient expression was also assumed in a report on *Agrobacterium*-mediated transformation of apple plants with green-fluorescent protein (Maximova et al. 1998). Also the kanamycin resistance *nptII* could be transiently expressed leading to a growing callus of untransformed plant cells. And additional an endogenous kanamycin tolerance was described for sweet orange plants (Peña et al. 1995) and the apple cultivar 'Pinova' was assumed to be more kanamycin tolerant than other *Malus* species (Flachowsky et al. 2008). If the kanamycin tolerance level can be high enough that untransformed calli could grow on selective media, was not reported. We did not test for *A. tumefaciens* contamination since the calli grew on a regeneration medium containing ticarcilin and cefotaxime which are antibiotics targeting *A. tumefaciens*. But the contamination with *A. tumefaciens* will be examined when putatively transgenic shoots are evaluated. The expression levels of *FB_MR5* could not be distinguished with regard to the regulatory elements controlling *FB_MR5* such as the CaMV 35S promoter, ocs terminator and the native promoter and terminator, respectively. The CaMV 35S promoter was described as a constitutively expressed promoter (Odell et al. 1985). The CaMV 35S promoter was also described to be regulated differently in different tissues of cotton as well as of soybean (Sunilkumar et al. 2002; Yang and Christou 2005). Therefor the expression levels of FB_MR5 in the ûndifferentiated callus tissue and in the leave tissue of MR5 are not assumed to be of equal strength and the different expression levels may not serve to draw a conclusion regarding the expression level in tissue derived from transgene lines. Summarized, an expression was found in more than 75% of sampled calli but expression levels may be different in transgene lines derived from these calli.

The overall results of the first completed part of the complementation assay pronounced that transformation is often successful but shoot regeneration remains as critical step to regenerate transgene plants. Therefore future work should emphasis on the evaluation of a more efficient method to induce shoot development with a series of different hormone derivatives and hormone dosages in plant shoot induction medium as well as different growth conditions. For the completion of the complementation assay regenerated shoots will be tested for target gene insertion and its expression, as well as the contamination with *A. tumefaciens* will be excluded by the amplification of a specific fragment with primers PicA1/PicA2 (Vanblaere 2011). The transgene *in vitro* shoots will be micro grafted in 12 replicates to M9 rootstocks and acclimatized to greenhouse conditions. These graftings will be inoculated with *E. amylovora* as described in Chapter 2.

References

Baumgartner, I. O., Patocchi, A., Franck, L., Kellerhals, M. and Broggini, G. A. L. (2011). "Fire blight resistance from 'Evereste' and *Malus sieversii* used in breeding for new high quality apple cultivars: Strategies and results." Acta Hort. (ISHS) 896: 391-397.

Belfanti, E., Silfverberg-Dilworth, E., Tartarini, S., Patocchi, A., Barbieri, M., Zhu, J., Vinatzer, B. A., Gianfranceschi, L., Gessler, C. and Sansavini, S. (2004). "The *HcrVf2* gene from a wild apple confers scab resistance to a transgenic cultivated variety." Proc. Natl. Acad. Sci. U. S. A. 101(3): 886-890.

Birch, R. G. (1997). "Plant transformation: Problems and strategies for practical application." Annu. Rev. Plant Physiol. Plant Mol. Biol. 48: 297-326.

Bondt, A., Eggermont, K., Penninckx, I., Goderis, I. and Broekaert, W. F. (1996). "*Agrobacterium* -mediated transformation of apple (*Malus x domestica* Borkh.): An assessment of factors affecting regeneration of transgenic plants." Plant Cell Rep. 15(7): 549-554.

Che, P., Lall, S., Nettleton, D. and Howell, S. H. (2006). "Gene expression programs during shoot, root, and callus development in *Arabidopsis* tissue culture." Plant Physiol. 141(2): 620-637.

Durel, C. E., Denance, C. and Brisset, M. N. (2009). "Two distinct major QTL for resistance to fire blight co-localize on linkage group 12 in apple genotypes 'Evereste' and *Malus floribunda* clone 821." Genome 52(2): 139-147.

Flachowsky, H., Le Roux, P.-M., Peil, A., Patocchi, A., Richter, K. and Hanke, M.-V. (2011). "Application of a high-speed breeding technology to apple (*Malus* x *domestica*) based on transgenic early flowering plants and marker-assisted selection." New Phytol. 192(2): 364-377.

Flachowsky, H., Riedel, M., Reim, S. and Hanke, M. (2008). "Evaluation of the uniformity and stability of T-DNA integration and gene expression in transgenic apple plants." Electronic Journal of Biotechnology 11(1).

Frey, J. E., Frey, B., Sauer, C. and Kellerhals, M. (2004). "Efficient low-cost DNA extraction and multiplex fluorescent PCR method for marker-assisted selection in breeding". Plant Breed. 123(6): 554-557.

Gamage, N. and Nakanishi, T. (2000). "*In vitro* shoot regeneration from leaf tissue of apple (cultiar "Orine"): Highshoot proliferation using carry over effect of TGZ." Acta Hort (ISHS) 520: 291-300.

Gessler, C. (2011). "Cisgenic disease resistant apples: A product with benefits for the environment, producer and consumer." Outlooks on Pest Management 22(5): 216-219.

Joshi, S. G. (2010). "Towards durable resistance to apple scab using cisgenes". Wageningen University, Wageningen, The Netherlands, PhD Thesis, Document number pages:

Le Roux, P.-M., Flachowsky, H., Hanke, M.-V., Gessler, C. and Patocchi, A. (2011). "Use of a transgenic early flowering approach in apple (*Malus* × *domestica* Borkh.) to introgress fire blight resistance from cultivar Evereste." Mol. Breed. 10.1007/s11032-011-9669-4: 1-18.

Liu, J. R., Sink, K. C. and Dennis, F. G. (1983). "Plant regeneration from apple seedling explants and callus cultures." Plant Cell, Tissue and Organ Culture 2(4): 293-304.

Malnoy, M. A., Korban, S., Boresjza-Wysocka, E. and Aldwinckle, H. C. (2008). "Apple". In "Compendium of transgenic crop plants: Transgenic temperate fruits and nuts". C. Kole and T. C. Hall. Hoboken, New Jersey, USA, Blackwell Publishing Ltd.

Maximova, S. N., Dandekar, A. M. and Guiltinan, M. J. (1998). "Investigation of *Agrobacterium*-mediated transformation of apple using green fluorescent protein: High transient expression and low stable transformation suggest that factors other than T-DNA transfer are rate-limiting." Plant Mol. Biol. 37(3): 549-559.

Milcevicova, R., Gosch, C., Halbwirth, H., Stich, K., Hanke, M. V., Peil, A., Flachowsky, H., Rozhon, W., Jonak, C., Oufir, M., Hausman, J. F., Matusikova, I., Fluch, S. and Wilhelm, E. (2010). "*Erwinia amylovora*-induced defense mechanisms of two apple species that differ in susceptibility to fire blight." Plant Sci 179(1-2): 60-67.

Norelli, J. L., Jones, A. L. and Aldwinckle, H. S. (2003). "Fire blight management in the twenty-first century: Using new technologies that enhance host resistance in apple." Plant Dis. 87(7): 756-765.

Norelli, J. L., Mills, J. Z., Timur Momol, M. and Aldwinckle, H. (1999). "Effect of cecropin-like transgenes on fire blight resistance of apple." Acta Hort (ISHS) 489: 273-278.

Odell, J. T., Nagy, F. and Chua, N.-H. (1985). "Identification of DNA sequences required for activity of the cauliflower mosaic virus 35S promoter." Nature 313(6005): 810-812.

Parravicini, G., Gessler, C., Denance, C., Lasserre-Zuber, P., Vergne, E., Brisset, M. N., Patocchi, A., Durel, C. E. and Broggini, G. A. L. (2011). "Identification of serine/threonine kinase and nucleotide-binding site-leucine-rich repeat (NBS-LRR) genes in the fire blight resistance quantitative trait locus of apple cultivar 'Evereste'." Mol. Plant Pathol. 12(5): 493-505.

Peil, A., Garcia-Libreros, T., Richter, K., Trognitz, F. C., Trognitz, B., Hanke, M. V. and Flachowsky, H. (2007). "Strong evidence for a fire blight resistance gene of *Malus robusta* located on linkage group 3." Plant Breed. 126(5): 470-475.

Peña, L., Cervera, M., Juárez, J., Navarro, A., Pina, J., Durán-Vila, N. and Navarro, L. (1995). "*Agrobacterium*-mediated transformation of sweet orange and regeneration of transgenic plants." Plant Cell Rep. 14(10): 616-619.

Schouten, H. J., Krens, F. A. and Jacobsen, E. (2006). "Cisgenic plants are similar to traditionally bred plants." EMBO Rep. 7(8): 750-753.

Sriskandarajah, S., Goodwin, P. B. and Speirs, J. (1994). "Genetic transformation of the apple scion cultivar 'Delicious' viaAgrobacterium tumefaciens." Plant Cell, Tissue and Organ Culture 36(3): 317-329.

Sunilkumar, G., Mohr, L., Lopata-Finch, E., Emani, C. and Rathore, K. S. (2002). "Developmental and tissue-specific expression of CaMV 35S promoter in cotton as revealed by GFP." Plant Mol. Biol. 50(3): 463-479.

Szankowski, I., Briviba, K., Fleschhut, J., Schönherr, J., Jacobsen, H. J. and Kiesecker, H. (2003). "Transformation of apple (*Malus domestica* Borkh.) with the stilbene synthase gene from grapevine (*Vitis vinifera* L.) and a PGIP gene from kiwi (*Actinidia deliciosa*)." Plant Cell Rep. 22(2): 141-149.

Szankowski, I., Waidmann, S., Degenhardt, J., Patocchi, A., Paris, R., Silfverberg-Dilworth, E., Broggini, G. A. L. and Gessler, C. (2009). "Highly scab-resistant transgenic apple lines achieved by introgression of *HcrVf2* controlled by different native promoter lengths." Tree Genetics & Genomes 5(2): 349-358.

Van der Hoorn, R. A. L. and Kamoun, S. (2008). "From guard to decoy: A new model for perception of plant pathogen effectors." Plant Cell 20(8): 2009-2017.

Vanblaere, T. (2011). "The development of a cisgenic scab resistant apple cv. 'Gala'". IBZ Plant Pathology, ETH Zurich, Doctor of Science, Document number pages: 122

Vanblaere, T., Szankowski, I., Schaart, J., Schouten, H., Flachowsky, H., Broggini, G. A. L. and Gessler, C. (2011). "The development of a cisgenic apple plant." J. Biotechnol. 154(4): 304-311.

Yang, N.-S. and Christou, P. (2005). "Cell type specific expression of a CaMV 35S-GUS gene in transgenic soybean plants." Dev. Genet. 11(4): 289-293.

Zhao, Y. F., He, S. Y. and Sundin, G. W. (2006). "The *Erwinia amylovora avrRpt2(EA)* gene contributes to virulence on pear and *AvrRpt2(EA)* is recognized by *Arabidopsis RPS2* when expressed in *Pseudomonas syringae*." Mol. Plant. Microbe Interact. 19(6): 644-654.

Chapter 5

General Conclusion and Discussion

Fire blight is a devastating disease representing enormous risks for apple producers to lose yield or even complete trees and orchards. The applied control measures such as copper or streptomycin need to be exactly timed in accordance to weather conditions otherwise the effectiveness is poor. Therefore producers are supported with forecasting tools like MaryBlyt™ but regardless the control of fire blight remains laborious and time consuming. The most efficacious measure to control the bacterium is given by the application of antibiotics. If the used compounds, namely streptomycin and oxytetracylcin, lose their effectiveness due to antibiotic resistant *E. amylovora* strains, new control measures need to be explored. These include the development of new antibiotics and new formulations of known antibiotics, such as kasugamycin, which have potential to replace streptomycin (McGhee and Sundin 2010; McGhee and Sundin 2011). Besides the latter an alternative measure could be the use of cultivars which are less susceptible. However, the known less susceptible cultivars do not have the necessary qualities to become popular. An alternative is to breed new, resistant cultivars by the introgression of natural resistances against *E. amylovora* occurring in wild and ornamental apple trees such as the strong resistances described for 'Evereste' on LG 12 (Durel et al. 2009; Parravicini et al. 2011) and *M.* × *robusta* 5 on LG 3 (Peil et al. 2007; Peil et al. 2008). To make the resistance usable for apple breeding and genetic modification we investigated the resistance locus of MR5 with the scope firstly to develop molecular markers usable for marker-assisted breeding and secondly to identify and characterize the gene(s) which deliver this resistance to the progeny.

The present work describes how in a first step the region of interest on LG 3 of MR5 comprising the fire blight locus was narrowed to 1,5 cM which may be approximately equal to 1,5 Mb (Velasco et al. 2010). In a second step, after increasing the mapping population to more than 2000 individuals, this region was enriched with 15 new developed or previously unmapped markers (SSRs and SNPs). After this, the region of interest defined by 33 recombinant individuals was limited by the markers FEM57 close to telomeric region of LG 3 and FEM18 towards centromere. The previously described most proximal marker 'Ch03e03' on LG 3 (Peil et al. 2007) was located in between FEM57 and FEM18 with a distance of 0.4 and 1.1 cM, respectively. In a third step the recombinants were assessed for their phenotype regarding resistance against *E. amylovora* strain Ea222_JKI. The collected PLL values transformed into binary values were used in a fourth step to map the locus between the flanking markers FEM14/FEM47 and rp16k15, flanking an interval of 0.23 cM. These molecular markers can be applied in marker-assisted breeding for fire blight resistance. To fortify the gained knowledge with the identification of a responsible resistance gene the region between the flanking markers was spanned in a fifth step by two bacterial artificial chromosomes harboring genomic DNA of MR5. One of them covered the region with DNA from the resistance giving chromosome; the second comprised the fragment from the homolog chromosome. Both BACs were sequenced and the sequence of the resistance carrying BAC was used for ORF prediction. The best evaluated protein prediction software was FGENESH which was trained on many organisms to predict mRNAs with specific splicing patterns (Salamov and Solovyev 2000; Yao et al. 2005). We predicted the ORFs with the algorithm trained on *Arabidopsis*, *V. vinifera* and tomato which were compared to each other. Most similarities were identified between the predictions of FGENESH trained on *Arabidopsis*/*Vitis* and *Arabidopsis*/tomato. Based on a comparison of the constitutional transcriptome of MR5 we suggest using the *Vitis* and tomato algorithms of FGENESH for gene prediction in apples. In a sixth step based on BLAST results and structural analyses of the predicted ORFs a single candidate resistance gene against FB, *FB_MR5*, was identified. Resequenced cDNA of *FB_MR5* confirmed the prediction with FGENESH trained on tomato. *FB_MR5* was shown to be constitutively expressed in MR5 and in tested resistant progenies of MR5. Tested susceptible cultivars and susceptible progenies of MR5 did not express *FB_MR5*. In a seventh step this gene was characterized *in silico* as a CNL comprising the *R* gene typical motifs and an LRR-like region. The present motifs of the NBS domain like Walker A and B, RNBS-B, MHD, etc. are indicators for the protein's functionality (Takken et al. 2006; Van Ooijen et al. 2008). The LRR-like region formed in a tertiary structure prediction a protein with a concave structure similar to a horse shoe like commonly known LRR proteins (Kobe and Deisenhofer 1994).

These findings indicated the functionality of *FB_MR5* which should be proven in a last step in which the candidate resistance gene was *Agrobacterium*-mediated integrated into the genome of susceptible 'Gala' plants. Results of the complementation assay are presented to the step of growing transgene shoots which are prepared for the production of replicates via micro graftings. This process was applied in apple (Joshi 2010) and used during the production process of cisgene apple plants (Vanblaere et al. 2011). Later, replicates of the transgene lines will be phenotyped by inoculation with *E. amylovora*.

Instead of the complementation of a susceptible cultivar another possibility would have been the silencing of *FB_MR5* in its original background of MR5. If an inoculation assay of such gene-silenced MR5 plant results in a susceptible phenotype, it is not finally proven that susceptibility is caused only by the silenced gene. It could be also possible, that other genes with sequence similarity to the target gene cause the complemented phenotype by 'cross-silencing'. Silencing "a block of *R* genes related by sequence similarity" with a single hairpin construct was demonstrated in transgenic *Nicotiana edwardsonii* which was silenced with a hairpin developed on an *R* gene homolog of *N. glutinosa* with approximately 83% similarity to *N*, an *R* gene against *Tobacco mosaic virus* (Balaji et al. 2007). Therefore we decided to evidence the functionality based on a complementation assay in susceptible 'Gala' apple plants. If the functionality will be proven positively, *FB_MR5* could be pyramided in e.g. cisgenic apple plants together with the putative FB resistance gene of 'Evereste' and additional scab and powdery mildew *R* genes such as *HcrVf2* and *Pl1/Pl2*, respectively. It could also be added to the genome of existing cisgene 'Gala' (Vanblaere et al. 2011) or it could be introduced into the multi-resistant early flowering trees (Le Roux et al. 2011). Pyramiding of *R* genes is often the strategy of choice especially if the target organisms reproduce clonally (McDonald and Linde 2002). Since the resistance of MR5 was broken in two described cases with Canadian *E. amylovora* strains (Norelli and Aldwinckle 1986; Peil et al. 2011) the pyramiding of several FB resistance genes is even more important and could contribute to a successful control strategy. But for example in case of *Xanthomonas campestris* pv. *vesicatoria* the strategy of pyramided *R* genes *Bs1*, *Bs2* and *Bs3* failed as reported for a pepper cultivar carrying all three *R* genes which were broken within one year (Kousik and Ritchie 1996). Putatively the breakdown of these pyramided *R* genes was only possible in such a short time because two of the corresponding Avr genes, *avrBs1* and *avrBs3*, were located on plasmids which were lost in new developed pathotypes (Kousik and Ritchie 1996). *AvrRpt2*$_{EA}$ is located on the chromosome. Therefore, a multi-resistant apple cultivar could become a key factor in the management of disease control in apple production and limit the use of fungicides and antibiotics.

The latter is of importance because antibiotics are widely used in agriculture especially in livestock but also in plant cultures mostly against fire blight and further for control of bacterial diseases of other high-value crops such as soft rot of cut flowers and potato caused by *Pectobacterium* spp., bacterial blight of celery caused by *Pseudomonas cichorii*, fruit-spotting or blossom blast symptoms of apple, pear and related trees caused by several pathovars of *P. syringae* and others (summarized in McManus et al. 2002). Regarding the possibility of occurring multidrug resistant human pathogenic bacteria and the beginning "era of untreatable infections" (Livermore 2009), there is significant need for human medicine to (1) slow down the evolution and selection of resistant strains and (2) to develop new antibiotics. The first task includes on the one hand the state of the art application of antibiotics in medicine and on the other hand the reduction of antibiotics used for animal livestock and fruit/vegetable cultivation. The second task seems to be disregarded and the development of new compounds failed. This is true especially for the development of antibacterial compounds functioning against Gram-negative bacteria (Arias and Murray 2009; Livermore 2009). *E. amylovora* is a Gram-negative bacterium. Therefore it is of special importance to reduce the application of streptomycin and oxytetracycline. The latter two antibiotics and active compounds of the same functional classes are also used in human medicine. The antibiotics are applied as a spray to the plants and may be later present in the soil, the groundwater, the plant, and the fruits but also in the secondary products such as honey. The antibiotic uptake by plants has been reported in recent publications (Kumar et al. 2005; Dolliver et al. 2007) and streptomycin has been detected in apples sprayed with this compound having concentrations of up to 18.4 µg/kg in the flesh and up to four times higher amounts in the apple skin (Mayerhofer et al. 2009). Therefore a possible uptake of antibiotics by consumers is not excluded which may lead to allergenic reactions or intoxication in humans and especially in children (Basaraba et al. 1999) which was described e.g. for beta-lactam antibiotics (Ponvert et al. 2011). Additional to the putatively occurring problems with antibiotic resistant *E. amylovora* strains spreading their resistance genes to human pathogenic bacteria, there is also the risk for apple producers to lose the only available and effective control measure to prevent fire blight infections in the field. Since streptomycin resistance is widespread in apple producing areas of North America and the alternative compounds are less effective, the need for new strategies in fighting fire blight grows. If the candidate *R* gene will be found to be functional and if it remains effective in other genetic backgrounds as it was in MR5, *FB_MR5* could be implemented in commercially important apple cultivars such as 'Fuji', 'Golden Delicious' or 'Gala' via cisgenesis and therefore the investigation is part of an innovative disease management with potential to limit the producers risk of high losses and to reduce the antibiotic applications. Applying several *R* genes pyramided in one marketable plant produced via cisgenesis (Gessler 2011) or early flowering (Le Roux et al. 2011) may become an additional tool in the orchestra of instruments against diseases and may lead to durable control measures in fighting FB. Furthermore, the achievement of resistant apple cultivars, which are marketable on the same level as its susceptible relatives ('Gala', 'Golden Delicious', 'Fuji', etc.), would bring an enormous reduction of work time and financial costs for apple producers as well as less environmental contamination.

R genes are usually found to be clustered (Meyers et al. 2003). Therefore we annotated the surrounding region (160 kb) of *FB_MR5*. All the predicted genes except the candidate *R* gene did not belong to classes of genes which were known to function as *R* genes. In the same workflow the question should be answered which of the prediction algorithms is the most promising in terms of accuracy. Therefore we compared the constitutional transcriptome of MR5 with the predicted ORFs. Results indicate a putative preference of the FGENESH algorithm trained on *V. vinifera* sequences instead of tomato and *Arabidopsis*. But for an evidence based answer the predicted transcripts need to be compared to separately amplified and resequenced transcripts or to RNA_Seq with higher coverage.

Future tasks on this research topic will comprise the verification of *FB_MR5* functional as *R* gene and the validation of the hypothetical mode of action of FB_MR5, RIN4_MR5 and AvrRpt2$_{EA}$. This work is already in preparation by our partners at JKI Dresden Pillnitz (Germany) who started yeast-2-hybrid experiments with FB_MR5/RIN4_MR5 and FB_MR5/AvrRpt2$_{EA}$. Parallel works should comprise the more technical task to create a multi-resistant apple cultivar. Due to most current directives of cisgenic apple cultivars (Gessler 2011) and the undefined legal status of cultivars produced with the 'early flowering' technology (Le Roux et al. 2011) also conventional breeding strategies should be used being as fast as possible to renew and give truth to the proverb “Ait a happle avore gwain to bed, An’ you’ll make the doctor beg his bread” (Wright 1913).

References

Arias, C. A. and Murray, B. E. (2009). "Antibiotic-resistant bugs in the 21st century — a clinical super-challenge". N. Engl. J. Med. 360(5): 439-443.

Balaji, B., Cawly, J., Angel, C., Zhang, Z., Palanichelvam, K., Cole, A. and Schoelz, J. (2007). "Silencing of the *N* family of resistance genes in *Nicotiana edwardsonii* compromises the hypersensitive response to tombusviruses". Mol. Plant. Microbe Interact. 20(10): 1262-1270.

Basaraba, R. J., Oehme, F. W., Vorhies, M. W. and Stokka, G. L. (1999). "Toxicosis in cattle from concurrent feeding of monensin and dried distiller's grains contaminated with macrolide antibiotics". J. Vet. Diagn. Invest. 11(1): 79-86.

Dolliver, H., Kumar, K. and Gupta, S. (2007). "Sulfamethazine uptake by plants from manure-amended soil". J. Environ. Qual. 36(4): 1224-1230.

Durel, C. E., Denance, C. and Brisset, M. N. (2009). "Two distinct major QTL for resistance to fire blight co-localize on linkage group 12 in apple genotypes 'Evereste' and *Malus floribunda* clone 821". Genome 52(2): 139-147.

Gessler, C. (2011). "Cisgenic disease resistant apples: A product with benefits for the environment, producer and consumer". Outlooks on Pest Management 22(5): 216-219.

Joshi, S. G. (2010). "Towards durable resistance to apple scab using cisgenes". Wageningen University, Wageningen, The Netherlands, PhD Thesis, Document number pages:

Kobe, B. and Deisenhofer, J. (1994). "The leucine-rich repeat: A versatile binding motif". Trends Biochem. Sci. 19(10): 415-421.

Kousik, C. S. and Ritchie, D. F. (1996). "Race shift in *Xanthomonas campestris* pv. *vesicatoria* within a season in field-grown pepper". Phytopathology 86(9): 952-958.

Kumar, K., Gupta, S. C., Baidoo, S. K., Chander, Y. and Rosen, C. J. (2005). "Antibiotic uptake by plants from soil fertilized with animal manure". J. Environ. Qual. 34(6): 2082-2085.

Le Roux, P.-M., Flachowsky, H., Hanke, M.-V., Gessler, C. and Patocchi, A. (2011). "Use of a transgenic early flowering approach in apple (*Malus* x *domestica* Borkh.) to introgress fire blight resistance from cultivar Evereste". Mol. Breed. 10.1007/s11032-011-9669-4: 1-18.

Livermore, D. M. (2009). "Has the era of untreatable infections arrived?". J. Antimicrob. Chemother. 64(suppl 1): i29-i36.

Mayerhofer, G., Schwaiger-Nemirova, I., Kuhn, T., Girsch, L. and Allerberger, F. (2009). "Detecting streptomycin in apples from orchards treated for fire blight". J. Antimicrob. Chemother. 63(5): 1076-1077.

McDonald, B. A. and Linde, C. (2002). "Pathogen population genetics, evolutionary potential, and durable resistance". Annu. Rev. Phytopathol. 40(1): 349-379.

McGhee, G. C. and Sundin, G. W. (2010). "Evaluation of kasugamycin for fire blight management, effect on nontarget bacteria, and assessment of kasugamycin resistance potential in *Erwinia amylovora*". Phytopathology 101(2): 192-204.

McGhee, G. C. and Sundin, G. W. (2011). "Kasumin: Field results for fire blight management and evaluation of the potential for spontaneous resistance development in *Erwinia amylovora* ". Acta Hort. (ISHS) 896: 519-525.

McManus, P. S., Stockwell, V. O., Sundin, G. W. and Jones, A. L. (2002). "Antibiotic use in plant agriculture". Annu. Rev. Phytopathol. 40: 443-+.

Meyers, B. C., Kozik, A., Griego, A., Kuang, H. and Michelmore, R. W. (2003). "Genome-wide analysis of NBS-LRR–encoding genes in *Arabidopsis*". Plant Cell 15(4): 809-834.

Norelli, J. L. and Aldwinckle, H. S. (1986). "Differential susceptibility of *Malus* spp cultivars Robusta-5, Novole, and Ottawa-523 to *Erwinia-Amylovora*". Plant Dis. 70(11): 1017-1019.

Parravicini, G., Gessler, C., Denance, C., Lasserre-Zuber, P., Vergne, E., Brisset, M. N., Patocchi, A., Durel, C. E. and Broggini, G. A. L. (2011). "Identification of serine/threonine kinase and nucleotide-binding site-leucine-rich repeat (NBS-LRR) genes in the fire blight resistance quantitative trait locus of apple cultivar 'Evereste'". Mol. Plant Pathol. 12(5): 493-505.

Peil, A., Flachowsky, H., Hanke, M.-V., Richter, K. and Rode, J. (2011). "Inoculation of *Malus* × *robusta* 5 progeny with a strain breaking resistance to fire blight reveals a minor QTL on LG5". Acta Hort. (ISHS) 896: 357-362.

Peil, A., Garcia-Libreros, T., Richter, K., Trognitz, F. C., Trognitz, B., Hanke, M. V. and Flachowsky, H. (2007). "Strong evidence for a fire blight resistance gene of *Malus robusta* located on linkage group 3". Plant Breed. 126(5): 470-475.

Peil, A., Hanke, M. V., Flachowsky, H., Richter, K., Garcia-Libreros, T., Celton, J. M., Gardiner, S., Horner, M. and Bus, V. (2008).

"Confirmation of the fire blight QTL of *Malus* x *robusta* 5 on linkage group 3". Acta Hort (ISHS) 793: 297-303.

Ponvert, C., Perrin, Y., Bados-Albiero, A., Le Bourgeois, M., Karila, C., Delacourt, C., scheinmann, P. and De Blic, J. (2011). "Allergy to betalactam antibiotics in children: results of a 20-year study based on clinical history, skin and challenge tests". Pediatr. Allergy Immunol. 22(4): 411-418.

Salamov, A. A. and Solovyev, V. V. (2000). "*Ab initio* gene finding in *Drosophila* genomic DNA". Genome Res. 10(4): 516-522.

Takken, F. L., Albrecht, M. and Tameling, W. I. (2006). "Resistance proteins: Molecular switches of plant defence". Curr. Opin. Plant Biol. 9(4): 383-390.

Van Ooijen, G., Mayr, G., Kasiem, M. M. A., Albrecht, M., Cornelissen, B. J. C. and Takken, F. L. W. (2008). "Structure–function analysis of the NB-ARC domain of plant disease resistance proteins". J. Exp. Bot. 59(6): 1383-1397.

Vanblaere, T., Szankowski, I., Schaart, J., Schouten, H., Flachowsky, H., Broggini, G. A. L. and Gessler, C. (2011). "The development of a cisgenic apple plant". J. Biotechnol. 154(4): 304-311.

Velasco, R., Zharkikh, A., Affourtit, J., Dhingra, A., Cestaro, A., Kalyanaraman, A., Fontana, P., Bhatnagar, S. K., Troggio, M., Pruss, D., Salvi, S., Pindo, M., Baldi, P., Castelletti, S., Cavaiuolo, M., Coppola, G., Costa, F., Cova, V., Dal Ri, A., Goremykin, V., Komjanc, M., Longhi, S., Magnago, P., Malacarne, G., Malnoy, M., Micheletti, D., Moretto, M., Perazzolli, M., Si-Ammour, A., Vezzulli, S., Zini, E., Eldredge, G., Fitzgerald, L. M., Gutin, N., Lanchbury, J., Macalma, T., Mitchell, J. T., Reid, J., Wardell, B., Kodira, C., Chen, Z., Desany, B., Niazi, F., Palmer, M., Koepke, T., Jiwan, D., Schaeffer, S., Krishnan, V., Wu, C., Chu, V. T., King, S. T., Vick, J., Tao, Q., Mraz, A., Stormo, A., Stormo, K., Bogden, R., Ederle, D., Stella, A., Vecchietti, A., Kater, M. M., Masiero, S., Lasserre, P., Lespinasse, Y., Allan, A. C., Bus, V., Chagne, D., Crowhurst, R. N., Gleave, A. P., Lavezzo, E., Fawcett, J. A., Proost, S., Rouze, P., Sterck, L., Toppo, S., Lazzari, B., Hellens, R. P., Durel, C.-E., Gutin, A., Bumgarner, R. E., Gardiner, S. E., Skolnick, M., Egholm, M., Van de Peer, Y., Salamini, F. and Viola, R. (2010). "The genome of the domesticated apple (*Malus* x *domestica* Borkh.)". Nat. Genet. 42(10): 833-839.

Wright, E. M. (1913). "Rustic Speech and Folk-Lore". Oxford, Oxford University Press.

Chapter 5

Yao, H., Guo, L., Fu, Y., Borsuk, L. A., Wen, T.-J., Skibbe, D. S., Cui, X., Scheffler, B. E., Cao, J., Emrich, S. J., Ashlock, D. A. and Schnable, P. S. (2005). "Evaluation of five *ab initio* gene prediction programs for the discovery of maize genes". Plant Mol. Biol. 57(3): 445-460.

Appendix

A, RNA-Seq mapping against predicted genes on LG 3 of MR5

All ORFs with at least one paired read (blue) matching 100% to the predicted mRNA. Green bar indicates the exon(s). In red and light green lines forward and reverse, respectively, reads of broken pairs are visualized. Reads were mapped with the RNA-Seq plugin of CLC Genomics Workbench 4.8. Colour picture at http://e-collection.library.ethz.ch/view/eth:6104

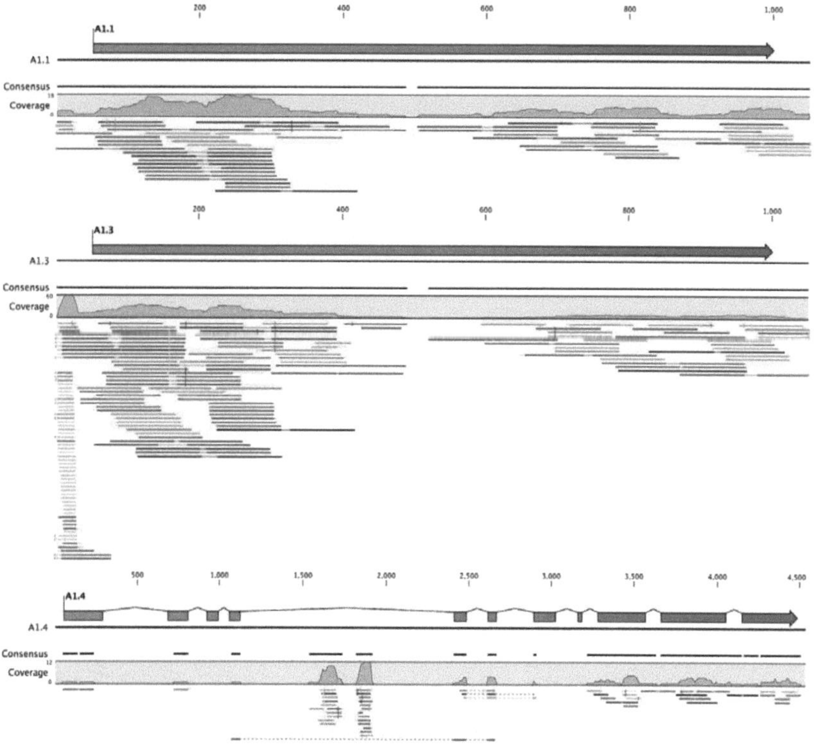

Appendix A

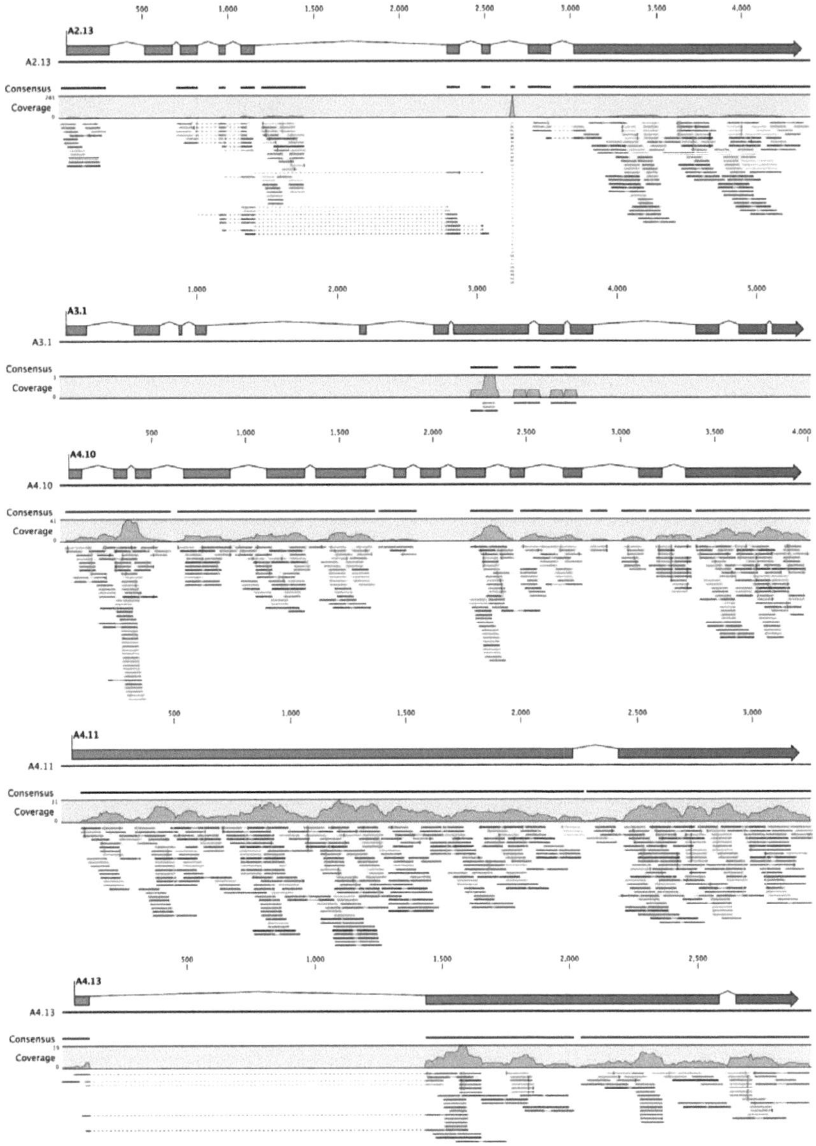
500
1,000
1,500
2,000
2,500
3,000
3,500
4,000
A2.13
A2.13
Consensus
Coverage
1,000
2,000
3,000
4,000
5,000
A3.1
A3.1
Consensus
Coverage
500
1,000
1,500
2,000
2,500
3,000
3,500
4,000
A4.10
A4.10
Consensus
Coverage
500
1,000
1,500
2,000
2,500
3,000
A4.11
A4.11
Consensus
Coverage
500
1,000
1,500
2,000
2,500
A4.13
A4.13
Consensus
Coverage

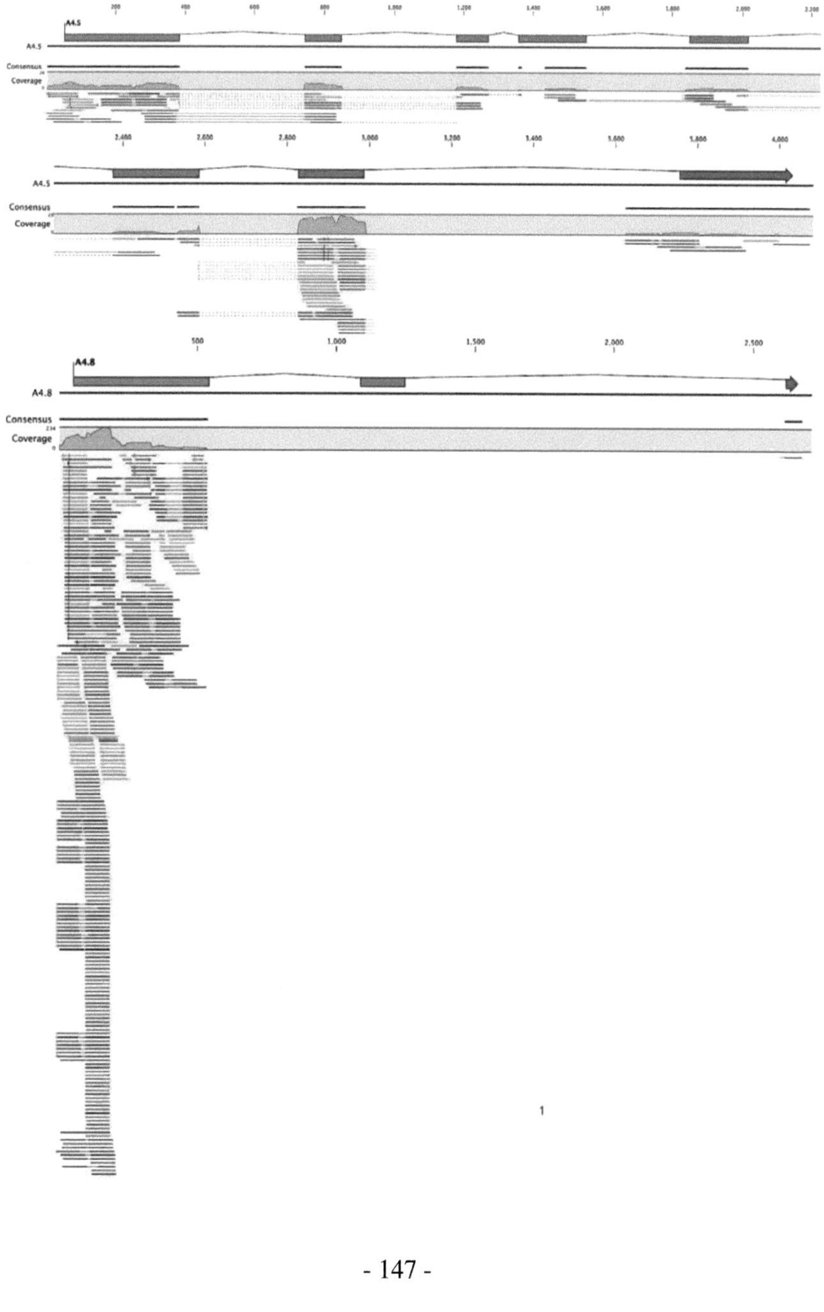
A4.5
Consensus
Coverage
A4.5
Consensus
Coverage
A4.8
A4.8
Consensus
Coverage
500
1,000
1,500
2,000
2,500

Appendix A

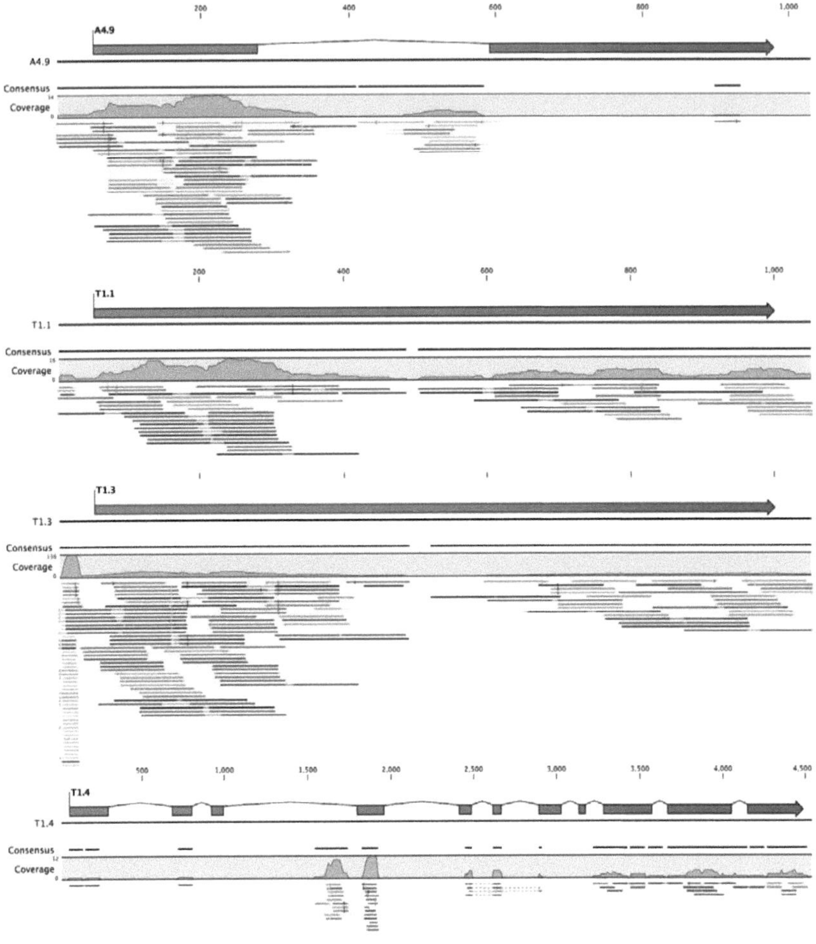

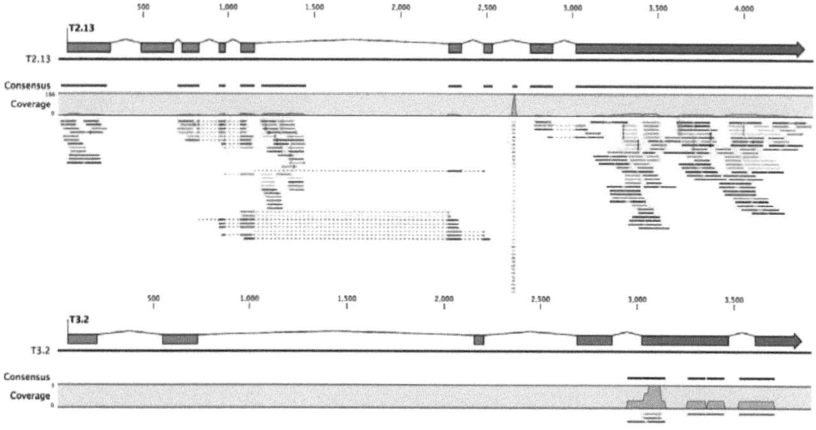
500
1,000
1,500
2,000
2,500
3,000
3,500
4,000
T2.13
T2.13
Consensus
Coverage
500
1,000
1,500
2,000
2,500
3,000
3,500
T3.2
T3.2
Consensus
Coverage

Appendix A

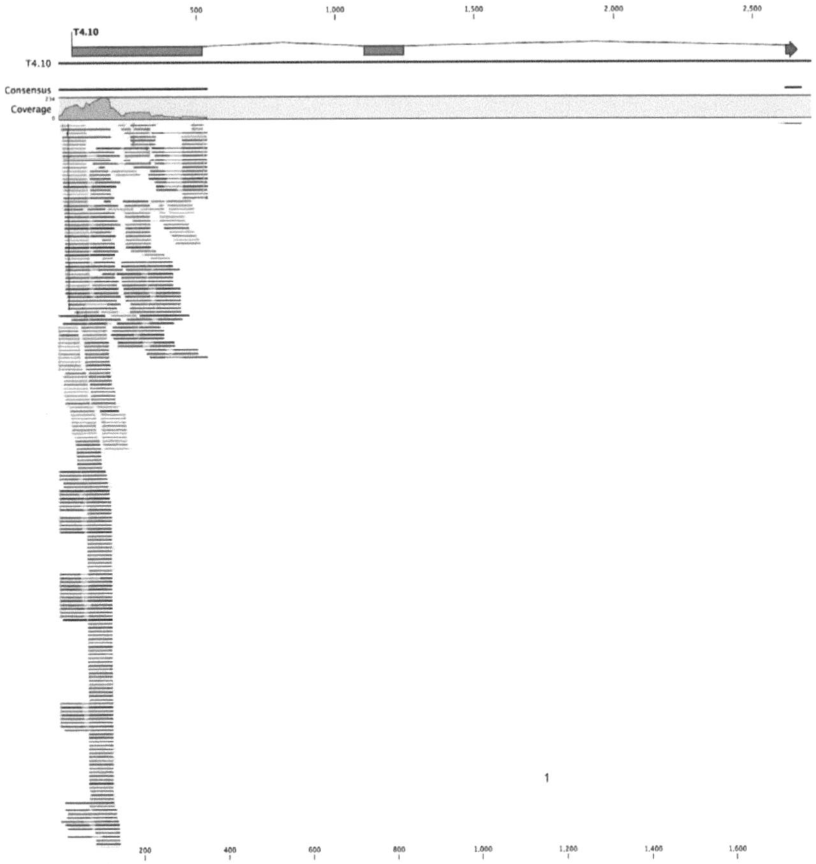
500
1,000
1,500
2,000
2,500
T4.10
T4.10
Consensus
Coverage

1

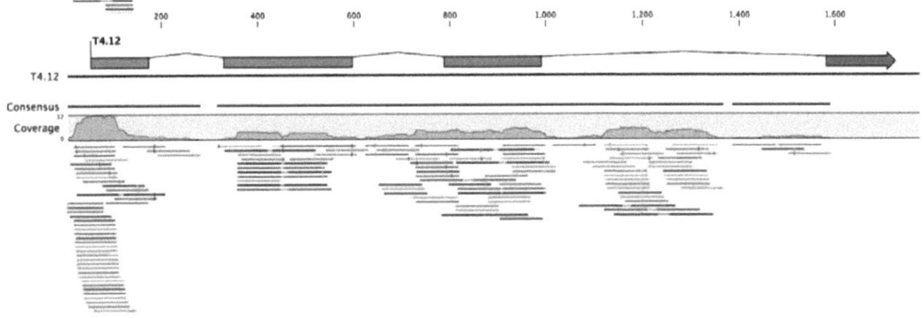
200
400
600
800
1,000
1,200
1,400
1,600
T4.12
T4.12
Consensus
Coverage

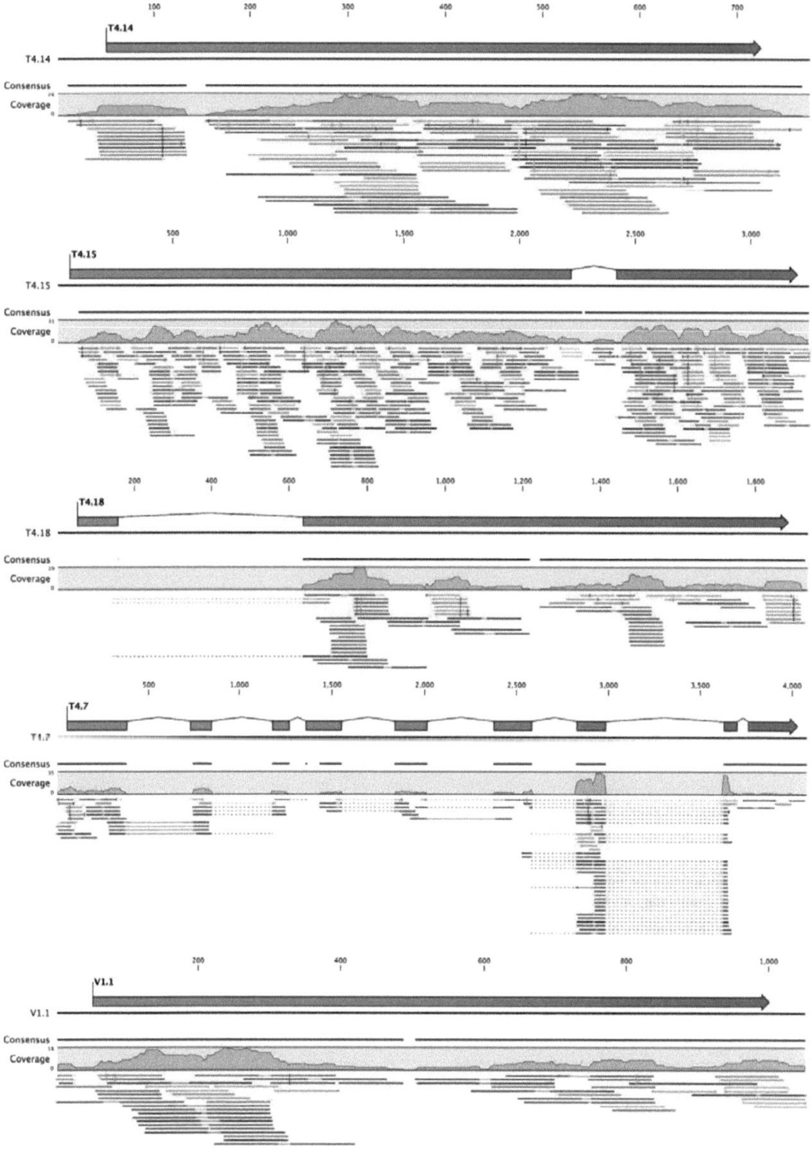
T4.14
Consensus
Coverage
T4.15
Consensus
Coverage
T4.18
Consensus
Coverage
T4.7
Consensus
Coverage
V1.1
Consensus
Coverage

Appendix A

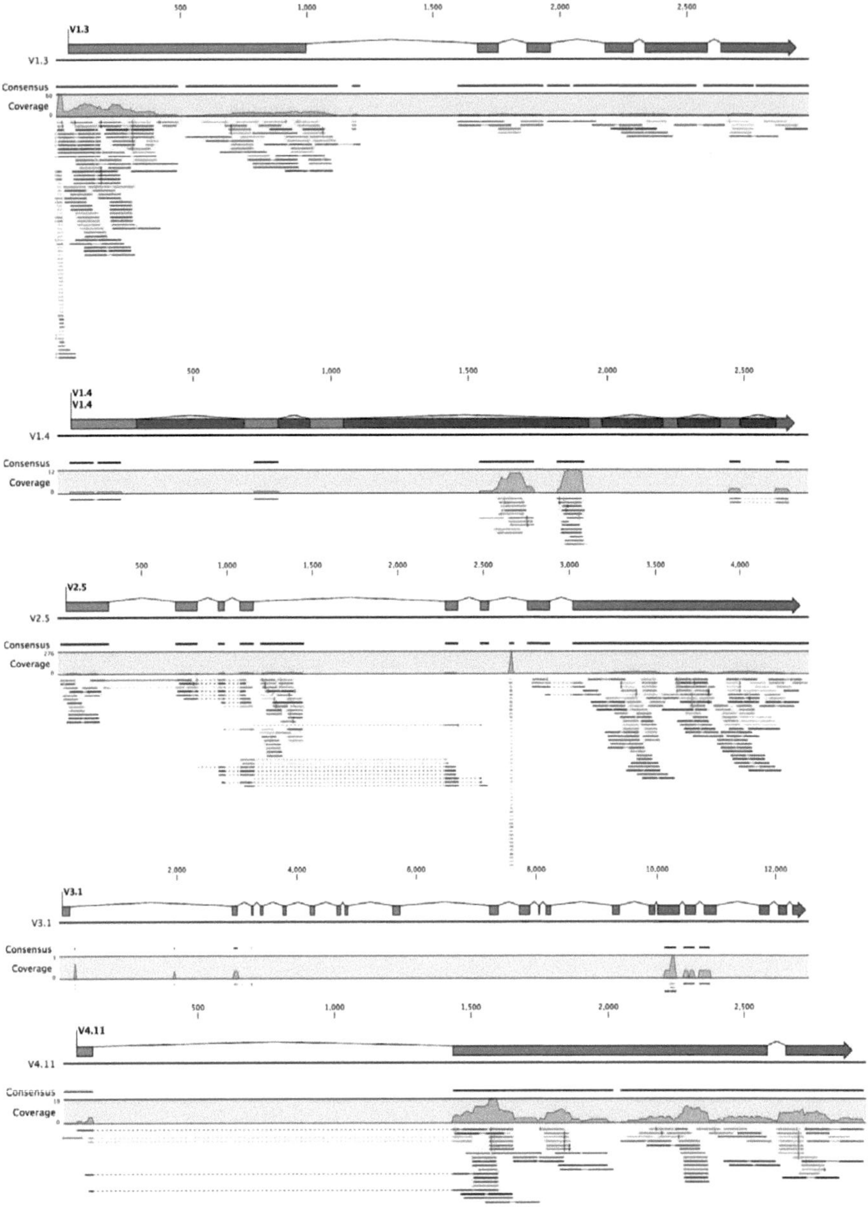
V1.3
Consensus
Coverage
500
1,000
1,500
2,000
2,500
V1.4
V2.5
3,000
3,500
4,000
V3.1
4,000
6,000
8,000
10,000
12,000
V4.11

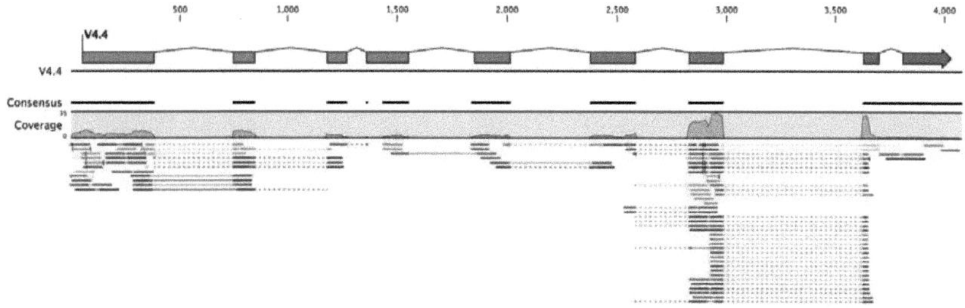
V4.4
Consensus
Coverage
500
1,000
1,500
2,000
2,500
3,000
3,500
4,000

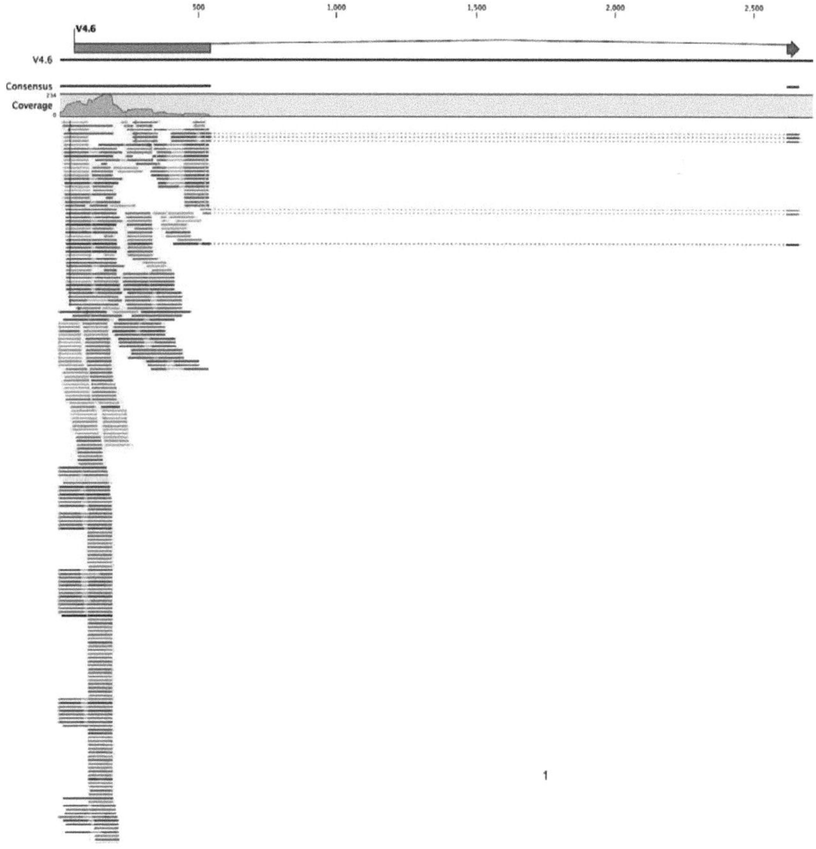
V4.6
Consensus
Coverage
500
1,000
1,500
2,000
2,500

1

Appendix A

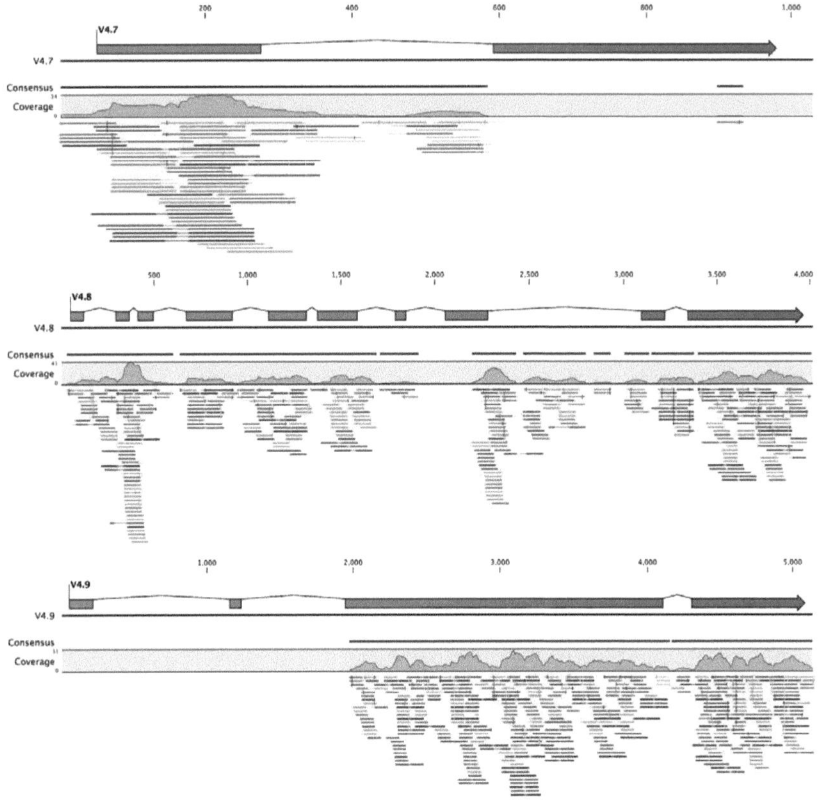
200
400
600
800
1,000
V4.7
V4.7
Consensus
Coverage
500
1,000
1,500
2,000
2,500
3,000
3,500
4,000
V4.8
V4.8
Consensus
Coverage
1,000
2,000
3,000
4,000
5,000
V4.9
V4.9
Consensus
Coverage

B, Nucleotide BLAST results of FGENESH predicted ORFs

Origin of sequences used to train FGENESH	Contig of BAC 16k15	ORF no.	ORF length	Accession number of BLASTn first hit	Description of BLASTn first hit	Score	E-value
V. vinifera	1	V1.1	951	XM_002271525	Predicted: *V.vinifera* uncharacterized	353	1E-93
V. vinifera	1	V1.2	1221	AM167520	*M. x domestica* transposon gene for putative DNA topoisomerase	1138	0
V. vinifera	1	V1.3	1791	no hits			
V. vinifera	1	V1.4	747	no hits			
V. vinifera	1	V1.5	330	no hits			
V. vinifera	1	V1.6	2085	AB270792	*M. x domestica* MdSFBB9-alpha, S9-RNase, MdSFBB9-beta genes	263	2E-66
V. vinifera	2	V2.1	369	HQ399004	*M. baccata* clone OLE1-19 NBS-LRR-like protein gene	196	2E-46
V. vinifera	2	V2.2	1236	no hits			
V. vinifera	2	V2.3*	4317	HQ399004	*M. baccata* clone OLE1-19 NBS-LRR-like protein gene	749	0
V. vinifera	2	V2.4	1473	no hits			
V. vinifera	2	V2.5	2115	no hits			
V. vinifera	2	V2.6	519	no hits			
V. vinifera	3	V3.1	2787	AM484137.2	*V. vinifera* contig VV78X060568.3	215	5E-52
V. vinifera	3	V3.2	456	no hits			
V. vinifera	4	V4.1	969	AM167520	*M. x domestica* transposon gene for putative DNA topoisomerase	248	5E-62
V. vinifera	4	V4.2	5112	DQ490951.2	*Zea mays* subsp. *mays* genotype CMS-S mitochondrion	58	1E-04
V. vinifera	4	V4.3	480	no hits			
V. vinifera	4	V4.4	1569	XM_002283399	Predicted: *V. vinifera* GPI mannosyltransferase 2-like	608	2E-170
V. vinifera	4	V4.5	582	AB627206	*M. x domestica* mRNA, microsatellite: MEST031	1048	0
V. vinifera	4	V4.6	540	no hits			
V. vinifera	4	V4.7	615	AM167520	*M. x domestica* transposon gene for putative DNA topoisomerase	396	2E-106
V. vinifera	4	V4.8	1953	AB270792	*M. x domestica* MdSFBB9-alpha, S9-RNase, MdSFBB9-beta genes	1094	0
V. vinifera	4	V4.9	3204	AM167520	*M. x domestica* transposon gene for putative DNA topoisomerase	3009	0
V. vinifera	4	V4.10	234	no hits			
V. vinifera	4	V4.11	1464	XM_002527015	*Ricinus communis* Protein AFR, putative	1088	0
Tomato	1	T1.1	951	XM_002271525	Predicted: *V. vinifera* uncharacterized	353	1E-93
Tomato	1	T1.2	1665	AM167520	*M. x domestica* transposon gene for putative DNA topoisomerase	1264	0
Tomato	1	T1.3	951	no hits			

Appendix B

Tomato	1	T1.4	1947	AB270792	*M. x domestica* MdSFBB9-alpha, S9-RNase, MdSFBB9-beta genes	207	9E-50
Tomato	1	T1.5	330	no hits			
Tomato	1	T1.6	402	no hits			
Tomato	1	T1.7	690	no hits			
Tomato	1	T1.8	600	AB270792	*M. x domestica* MdSFBB9-alpha, S9-RNase, MdSFBB9-beta genes	241	9E-60
Tomato	1	T1.9	480	AB270792	*M. x domestica* MdSFBB9-alpha, S9-RNase, MdSFBB9-beta genes	272	3E-69
Tomato	2	T2.1	669	no hits			
Tomato	2	T2.2	531	HQ399004	*M. baccata* clone OLE1-19 NBS-LRR-like protein gene	196	2E-46
Tomato	2	T2.3	609	no hits			
Tomato	2	T2.4	480	no hits			
Tomato	2	T2.5	198	no hits			
Tomato	2	T2.6	843	no hits			
Tomato	2	T2.7*	4167	HQ399004	*M. baccata* clone OLE1-19 NBS-LRR-like protein gene	749	0
Tomato	2	T2.8	1278	no hits			
Tomato	2	T2.9	519	no hits			
Tomato	2	T2.10	987	no hits			
Tomato	2	T2.11	696	no hits			
Tomato	2	T2.12	177	no hits			
Tomato	2	T2.13	2283	AM462858	*V. vinifera*, whole genome shotgun sequence, contig VV78X147076.10	100	2E-17
Tomato	2	T2.14	480	no hits			
Tomato	3	T3.1	519	no hits			
Tomato	3	T3.2	1275	AM484137.2	*V. vinifera* contig VV78X060568.3	180	2E-41
Tomato	3	T3.3	528	no hits			
Tomato	3	T3.4	447	AY603367	*M. x domestica* isolate 2 retrotransposon Cassandra	183	2E-42
Tomato	3	T3.5	1707	no hits			
Tomato	3	T3.6	291	no hits			
Tomato	4	T4.1	1098	XM_002527735	*Ricinus communis* conserved hypothetical protein	150	2E-32
Tomato	4	T4.2	249	no hits			
Tomato	4	T4.3	420	no hits			
Tomato	4	T4.4	120	no hits			
Tomato	4	T4.5	5013	DQ490951.2	*Zea mays* subsp. *mays* genotype CMS-S mitochondrion	58	1E-04
Tomato	4	T4.6	480	no hits			
Tomato	4	T4.7	1647	XM_002283399	Predicted: *V. vinifera* GPI mannosyltransferase 2-like	606	8E-170
Tomato	4	T4.8	261	no hits			
Tomato	4	T4.9	582	AB627206	*M. x domestica* mRNA, microsatellite: MEST031	1048	0
Tomato	4	T4.10	663	AB622414	*Menispermum dauricum* DNA	64	2E-06
Tomato	4	T4.11	471	EU794445	*M. floribunda* clone M18S-5Cs Vf apple scab resistance protein	147	2E-31

					HcrVf2-like gene		
Tomato	4	T4.12	747	EU794445	*M. floribunda* clone M18S-5Cs Vf apple scab resistance protein HcrVf2-like gene	457	8E-125
Tomato	4	T4.13	171	AB270792	*M. x domestica* MdSFBB9-alpha, S9-RNase, MdSFBB9-beta genes	283	1E-72
Tomato	4	T4.14	675	AB270792	*M. x domestica* MdSFBB9-alpha, S9-RNase, MdSFBB9-beta genes	1182	0
Tomato	4	T4.15	2961	AM167520	*M. x domestica* transposon gene for putative DNA topoisomerase	3009	0
Tomato	4	T4.16	405	no hits			
Tomato	4	T4.17	234	no hits			
Tomato	4	T4.18	1359	XM_002297609	*Populus trichocarpa* f-box family protein	1149	0
Arabidopsis	1	A1.1	951	XM_002271525	Predicted: *V. vinifera* uncharacterized	353	1E-93
Arabidopsis	1	A1.2	933	AM167520	*M. x domestica* transposon gene for putative DNA topoisomerase	1173	0
Arabidopsis	1	A1.3	951	no hits			
Arabidopsis	1	A1.4	1833	no hits			
Arabidopsis	1	A1.5	330	no hits			
Arabidopsis	1	A1.6	501	no hits			
Arabidopsis	1	A1.7	1740	no hits			
Arabidopsis	1	A1.8	432	AB270792	*M. x domestica* MdSFBB9-alpha, S9-RNase, MdSFBB9-beta genes	361	7E-96
Arabidopsis	1	A1.9	792	AB270792	*M. x domestica* MdSFBB9-alpha, S9-RNase, MdSFBB9-beta genes	560	6E-156
Arabidopsis	2	A2.1	249	no hits			
Arabidopsis	2	A2.2	369	HQ399004	*M. baccata* clone OLE1-19 NBS-LRR-like protein gene	196	2E-46
Arabidopsis	2	A2.3	276	no hits			
Arabidopsis	2	A2.4	528	no hits			
Arabidopsis	2	A2.5	198	no hits			
Arabidopsis	2	A2.6	816	no hits			
Arabidopsis	2	A2.7*	4317	HQ399004	*M. baccata* clone OLE1-19 NBS-LRR-like protein gene	749	0
Arabidopsis	2	A2.8	1314	no hits			
Arabidopsis	2	A2.9	531	no hits			
Arabidopsis	2	A2.10	546	no hits			
Arabidopsis	2	A2.11	705	no hits			
Arabidopsis	2	A2.12	300	no hits			
Arabidopsis	2	A2.13	2256	AM462858	*V. vinifera*, whole genome shotgun sequence, contig VV78X147076.10	100	2E-17
Arabidopsis	2	A2.14	519	no hits			
Arabidopsis	3	A3.1	2118	AM484137.2	*V. vinifera* contig VV78X060568.3	217	2E-52
Arabidopsis	3	A3.2	528	no hits			
Arabidopsis	3	A3.3	708	no hits			
Arabidopsis	3	A3.4	516	no hits			

Appendix B

Arabidopsis	3	A3.5	843	no hits			
Arabidopsis	4	A4.1	885	XM_002283399	Predicted: *V. vinifera* GPI mannosyltransferase 2-like	268	4E-68
Arabidopsis	4	A4.2	507	no hits			
Arabidopsis	4	A4.3	5112	DQ490951.2	*Zea mays* subsp. *mays* genotype CMS-S mitochondrion	58	1E-04
Arabidopsis	4	A4.4	480	no hits			
Arabidopsis	4	A4.5	1545	XM_002283399	Predicted: *V. vinifera* GPI mannosyltransferase 2-like	608	2E-170
Arabidopsis	4	A4.6	255	no hits			
Arabidopsis	4	A4.7	582	AB627206	*M. x domestica* mRNA, microsatellite: MEST031	1048	0
Arabidopsis	4	A4.8	702	AB622414	*Menispermum dauricum* DNA	64	2E-06
Arabidopsis	4	A4.9	615	AM167520	*M. x domestica* transposon gene for putative DNA topoisomerase	396	2E-106
Arabidopsis	4	A4.10	2244	AB270792	*M. x domestica* MdSFBB9-alpha, S9-RNase, MdSFBB9-beta genes	1094	0
Arabidopsis	4	A4.11	2961	AM167520	*M. x domestica* transposon gene for putative DNA topoisomerase	3009	0
Arabidopsis	4	A4.12	234	no hits			
Arabidopsis	4	A4.13	1464	XM_002527015	*Ricinus communis* Protein AFR, putative	1088	0

Acknowledgments

I kindly acknowledge my examiners

CG, BMcD and AP.

I further thank MK who leaded the project ZUEFOS and I am thankful for financial support from Federal Office for Agriculture FOAG of Switzerland (project: ZUEFOS) as well as the D-A-CH (project: 310030L_130811).

Further I like to thank all members of the Plant Pathology Group from IBZ (ETHZ) and especially GB. I thank the colleagues from ACW Wädenswil, JKI Dresden-Pillnitz and Quedlinburg as well as the colleagues from GDC (ETHZ) who supported me during the last three years.

Special thanks go to my family who moved to Zurich with me and supported me during the entire time.

MIX
Papier aus verantwortungsvollen Quellen
Paper from responsible sources
FSC® C105338

Printed by Books on Demand GmbH, Norderstedt / Germany